AF765191

Monoclonal Antibodies

Physicochemical Analysis

Monoclonal Antibodies
Physicochemical Analysis

Edited by

HARLEEN KAUR
Biologics Division, Aurobindo Pharma Ltd., Hyderabad, India; Aurobindo Biologics (Unit-XVII) (CurateQ Biologics), Hyderabad, India

DIETMAR REUSCH
Pharma Technical Development, Roche Diagnostics GmbH, Penzberg, Germany

ELSEVIER

ACADEMIC PRESS
An imprint of Elsevier

Academic Press is an imprint of Elsevier
125 London Wall, London EC2Y 5AS, United Kingdom
525 B Street, Suite 1650, San Diego, CA 92101, United States
50 Hampshire Street, 5th Floor, Cambridge, MA 02139, United States
The Boulevard, Langford Lane, Kidlington, Oxford OX5 1GB, United Kingdom

Copyright © 2021 Harleen Kaur and Dietmar Reusch. Published by Elsevier Inc.
All Rights Reserved.

No part of this publication may be reproduced or transmitted in any form or by any means, electronic or mechanical, including photocopying, recording, or any information storage and retrieval system, without permission in writing from the publisher. Details on how to seek permission, further information about the Publisher's permissions policies and our arrangements with organizations such as the Copyright Clearance Center and the Copyright Licensing Agency, can be found at our website: www.elsevier.com/permissions.

This book and the individual contributions contained in it are protected under copyright by the Publisher (other than as may be noted herein).

Notices
Knowledge and best practice in this field are constantly changing. As new research and experience broaden our understanding, changes in research methods, professional practices, or medical treatment may become necessary.

Practitioners and researchers must always rely on their own experience and knowledge in evaluating and using any information, methods, compounds, or experiments described herein. In using such information or methods they should be mindful of their own safety and the safety of others, including parties for whom they have a professional responsibility.

To the fullest extent of the law, neither the Publisher nor the authors, contributors, or editors, assume any liability for any injury and/or damage to persons or property as a matter of products liability, negligence or otherwise, or from any use or operation of any methods, products, instructions, or ideas contained in the material herein.

British Library Cataloguing-in-Publication Data
A catalogue record for this book is available from the British Library

Library of Congress Cataloging-in-Publication Data
A catalog record for this book is available from the Library of Congress

ISBN: 978-0-12-822318-5

For Information on all Academic Press publications
visit our website at https://www.elsevier.com/books-and-journals

Publisher: Andre Gerhard Wolff
Acquisitions Editor: Glyn Jones
Editorial Project Manager: Samantha Allard
Production Project Manager: Selvaraj Raviraj
Cover Designer: Miles Hitchen

Typeset by MPS Limited, Chennai, India

Working together
to grow libraries in
developing countries
www.elsevier.com • www.bookaid.org

Dedication

Dr. Harleen Kaur would like to dedicate this book to her mother (Kanwaljit Kaur) for having faith in her and always encouraging her.

Contents

List of contributors

Lowell J. Brady
Drug Product Analytical Development, Drug Product Services, Lonza AG, Basel, Switzerland

Patrick Bulau
Pharma Technical Development, Roche Diagnostics GmbH, Penzberg, Germany

Yan Chen
Pharma Technical Development, Genentech Inc, South San Francisco, CA, United States

Chris Chumsae
Bristol Myers Squibb, Process Development Analytics, Devens, MA, United States

Thomas Flad
Protagen Protein Services GmbH, Heilbronn, Germany

Tobias Graf
Pharma Technical Development, Roche Diagnostics GmbH, Penzberg, Germany

Christoph Gstöttner
Leiden University Medical Center, Center for Proteomics and Metabolomics, The Netherlands

Markus Haberger
Pharma Technical Development, Roche Diagnostics GmbH, Penzberg, Germany

Harleen Kaur
Analytical Sciences, Aurobindo Biologics (Unit-XVII) (CurateQ Biologics), Hyderabad, India

Michael Leiss
Pharma Technical Development, Roche Diagnostics GmbH, Penzberg, Germany

Manfred Wuhrer
Leiden University Medical Center, Center for Proteomics and Metabolomics, The Netherlands

Kai Zheng
Pharmaceutical Development, Genentech Inc, South San Francisco, CA, United States

Preface

This book provides an in-depth overview of the monoclonal antibody (mAb) structure and analytical methods that are routinely employed for the physicochemical characterization of mAbs. Specifically, the book discusses in detail the various quality attributes assessed as part of the physiochemical testing in the biopharmaceutical industry.

Dr. Harleen Kaur, the lead editor of this book, provides an introductory chapter (Chapter 1, Overview of Monoclonal Antibodies) describing the promising potential of the mAbs and gives an overview of mAbs. She has highlighted the history of mAbs and different theories that were proposed related to the existence and production of mAbs. In addition, the chapter explains the molecular structure of antibodies and expanded on the Fab (fragment, antigen-binding region), the Fc (fragment, crystallizable) region, the variable domains (VH and VL), and constant domains (CH1, CH2, CH3, and CL) of heavy and light chains in the antibody and their effect on antibody therapeutic mechanism. A section on different kinds of FDA-approved mAbs such as murine, humanized, and human have been presented along with their therapeutic applications.

Dr. Harleen Kaur, the lead editor of this book, is the author of Chapter 2, Physicochemical Characterization of Monoclonal Antibodies, which covers the importance of the physicochemical characterization of mAbs for regulatory filing. Elaborating on the antibody characterization, it is important to assess the critical quality attributes that impact the immunogenicity/safety, biological activity/efficacy, clearance (pharmacokinetics and pharmacodynamics, PK/PD), and overall structure and function of mAbs. In general, mAbs are commonly characterized for size, purity, identity, glycan profiling, and higher-order structure (HOS) as part of the batch release and stability testing.

Glycosylation is one of the most common posttranslational modifications (PTMs) observed in the mAbs. In Chapter 3, Glycosylation Analysis, Christoph Gstoettner, Dr. Harleen Kaur, and Dr. Manfred Wuhrer present the impact of different glycoforms such as galactose, sialic acid, and fucose on the biological behavior of the mAbs. The authors review the different analytical approaches adopted by the researchers to characterize the *N*-glycosylation and *O*-glycosylation in mAbs.

In Chapter 4, Primary Structure and Peptide Mapping, the peptide mapping method, which serves as an identity test to determine the primary structure of mAbs, is presented in detail by Drs. Thomas Flad and Chris Chumsae. The commonly observed post- and co-translational modifications (PTMs/CTMs) in mAbs have been enumerated. Mass spectrometry (MS) in conjunction with liquid chromatography (LC) also referred to as LC/MS or LC/MS/MS is a critical tool to determine the amino acid sequence and determine the co-translational (glycosylation) and post-translational modifications such as oxidation, deamidation, and isomerization. The authors have elaborated extensively on the use of MS for peptide mapping, the designing of mobile phase gradient, selection of ionization techniques and mass detectors. Additionally, the application of multimode attribute approach MS for assessing various quality attributes at different stages of the drug development has been considered.

In Chapter 5, Separation Techniques for Characterization and Quantification of Antibody Purity, Lowell J. Brady provides a comprehensive review of separation-based analytical techniques employed for quantification of antibody purity and product-related variants. The principle, working, and advantages and disadvantages of size and charge-based separation methods, such as size-exclusion chromatography, cation-exchange high-performance liquid chromatography, isoelectric focusing, and capillary zone electrophoresis have been addressed in depth. This chapter will also discuss the orthogonal techniques that are product-specific or required in the early stages of drug development, such as hydrophobic interaction chromatography and Boronate affinity chromatography.

Determination of antibody HOS by various analytical techniques is subjected to detail in Chapter 6, Higher Order Structure and Free Thiol, by Drs. Yan Chen and Kai Zheng. The structural integrity of the protein is critical for its biological activity and stability and henceforth, characterization of protein structure including secondary, tertiary, and quaternary structure is an important aspect of the physicochemical characterization. The HOS of mAb is demonstrated by conventional spectroscopic methods, such as nuclear magnetic resonance, X-ray crystallography, circular dichroism, Fourier-transform infrared spectroscopy, and fluorescence spectroscopy. In this chapter, the authors have explained the capabilities and limitations of each of the techniques with respect to its ability to monitor different aspects of the protein structure. The determination of free thiol (or sulfhydryl group) content in antibodies round out the discussion of the chapter.

In Chapter 7, Advanced Structural Elucidation of New Antibody Formats by Mass Spectrometry, Tobias Graf, Markus Haberger, Dr. Michael Leiss, and Dr. Patrick Bulau outline the need for emerging LC-MS-based analytical technologies while considering the development of the complex novel antibody formats, such as bispecific antibodies. The authors discuss the key role of the MS technique for antibody characterization and identification of modifications. Various LC-MS studies have been mentioned in detail to assess the size variants, charge heterogeneity, and hydrophobicity in mAbs. In addition, the authors assert the use of 2D-LC approaches for the characterization of new bio-therapeutic formats at the native protein level.

Acknowledgment

The editor is thankful to all the scientists who have contributed to this book in times of COVID-19 pandemic.

Harleen Kaur and Dietmar Reusch

CHAPTER 1

Overview of monoclonal antibodies

Harleen Kaur
Analytical Sciences, Aurobindo Biologics (Unit-XVII) (CurateQ Biologics), Hyderabad, India

1.1 Introduction

Antibodies also referred to as immunoglobulins (Ig) are large Y-shaped soluble proteins naturally produced by the immune system in response to the invading foreign particles or abnormal agents (antigens) such as harmful pathogens and viruses. Antibodies are one of the first lines of defense in higher organisms that recognize and bind these antigens to remove them from the body by phagocytosis or complement lysis. The part of the antibody that binds with antigens is known as paratope (or antigen-binding region) or complementarity-determining regions (CDRs). The paratope is located on the variable region of both the antibody's heavy chain (HC) and light chain (LC). The part of the antigen that is recognized by the antibody is referred to as epitope or antigenic determinant. With the antibody's ability to bind with high affinity and high specificity to the antigen surface, they have been widely explored as therapeutic and diagnostic tools for the treatment of diseases and the detection of antigens (analyte). Although very successful in targeting pathogens and increasingly being used in the clinic for treatment of several diseases for few decades, they are still compared against their rival molecules termed as "Aptamers" or often called as "chemical antibodies" in therapeutics and diagnostics industry (Kaur & Yung, 2012; Kaur, Li, Bay, & Yung, 2013; Kaur, Bruno, Kumar, & Sharma, 2018). Aptamers exhibit significant advantages over the antibodies in terms of small size, nonimmunogenic nature, less batch to batch variability and sensitivity to temperature and pH. However, with antibodies having roots in the immune system and extensive information already known about their structure, binding, and function, for the time being therapeutic antibodies are heavily dominating the biologics therapeutic market and are expected to grow at the compound annual growth rate of around 12.5% between 2017 and 2023 with

Monoclonal Antibodies
DOI: https://doi.org/10.1016/B978-0-12-822318-5.00006-5
Copyright © 2021 Harleen Kaur.
Published by Elsevier Inc. All rights reserved.

revenue generation of USD 218.97 billion by the end of 2023 (https://www.globenewswire.com/news-release/2018/04/10/1467446/0/en/Global-Monoclonal-Antibody-Therapeutics-Market-Will-Reach-USD-218-97-Billion-by-2023-Zion-Market-Research.html).

Antibodies may be polyclonal or monoclonal depending on their distinct properties and the way they are generated (Pohanka, 2009). Polyclonal antibodies are heterogeneous mixture of Igs produced by different clones of B-cells against a specific antigen but can bind to different epitopes of the same antigen. On the other hand, monoclonal antibodies (mAbs) are homogenous population of antibodies that are produced from single clone of B-cells and interact with specific epitope on the antigen surface. Derivation from a single clone of cells and subsequently targeting single epitope on the antigen surface is what differentiates mAbs from polyclonal antibodies. The high specificity of the mAb for an epitope reduces the possibility of the cross-reactivity, batch to batch variability, allows to conduct structural studies to understand the conformational changes, and provides concise and accurate quantitative analytical assay results which may not be the case for the naturally producing polyclonal antibodies. However, compared to polyclonal antibodies, the process to produce mAbs is expensive, laborious, and time consuming, a drawback that needs to be overcome.

Over the last few decades, the rise in the applications of mAbs has been remarkable. They have become important tools in basic research and have taken center stage in the area of therapeutics and diagnostics. The therapeutic market for mAb has witnessed the accelerated growth in the biopharmaceutical sector with over 80 mAbs granted marketing approval by regulatory agencies for treatment of several disorders like oncology, respiratory diseases, autoimmune, and inflammatory diseases. In addition, the introduction of biosimilars aimed at decreasing the cost of medication while increasing patient access to healthcare presents an attractive opportunity which has boosted the market growth in emerging and developed countries.

This chapter titled "Overview of Monoclonal Antibodies" is divided into three sections—the first section (Section 1.1) will provide an overview of the historical milestones in the discovery of antibodies, and the second section (Section 1.2) will present the chemical structure of antibodies with special focus on the IgG structure which is the most abundant antibody (75%) in body fluids and the most approved Ig as a therapeutic drug. The relative abundance of IgG in serum is $8-16\ \mathrm{mg\ mL^{-1}}$. Finally, the third section (Section 1.3) will highlight the importance of mAbs in biopharmaceutical industry and their therapeutic applications.

1.2 Discovery of monoclonal antibodies

The discovery of the antibodies dates back to 1890s when Emil Adolf von Behring and Kitasato Shibasaburōs work into "serum therapy" showed that serum from an animal immunized with diphtheria or tetanus toxin could be used to neutralize the effect of fatal doses of toxin in another animal (https://www.globenewswire.com/news-release/2018/04/10/1467446/0/en/Global-Monoclonal-Antibody-Therapeutics-Market-Will-Reach-USD-218-97-Billion-by-2023-Zion-Market-Research.html; Pohanka, 2009). This discovery was one of the remarkable developments in the field of medicine at the time that saved many lives and opened up new therapeutic possibilities. Behring and Kitasato also introduced the terminology "antitoxic" for the serum component present in the animals that were immunized with bacterial toxins and acquired immunity against the toxin (Behring & Kitasato, 1890; Grundbacher, 1992). The antitoxins are commonly referred to as antibodies now. Behring won the Nobel Prize in Physiology or Medicine in 1901 for his work on serum therapy particularly the discovery of a diphtheria antitoxin (https://www.nobelprize.org/prizes/medicine/1901/summary/). In 1897 bacteriologist Paul-Ehrlich proposed the side-chain theory of toxicity which became basis of immunological research at the time. This theory postulated that cells have side chains present on their surface for the adsorbance of the nutrients. These side chains can also bind to foreign substances like toxins (diphtheria/tetanus toxin or microorganism) in a manner similar to lock and key model for enzymes and their substrates (Bosch & Rosich, 2008; Winau, Westphal, & Winau, 2004). A large number of these side chains would be released into the bloodstream and act as antibodies or antitoxins to neutralize the toxin and thus preventing the binding of the toxin to other toxins in the blood. In the following years, Ehrlich extended his side-chain research and imagined side chains to be receptives that bind to toxins and introduced the term "receptor." Later, he collaborated with John Newport Langley who was investigating the effects of the alkaloids on muscle cells and nerves, and together they developed the receptor theory and extended the concept of receptor to "chemoreceptor" to understand the interaction of the drugs with the cells (Bennett, 2000). Ehrlich also contributed to the field of cancer research and introduced chemotherapy for the treatment of trypanosoma infections. In 1908 Nobel Prize in Physiology or Medicine was awarded to Paul-Ehrlich jointly with Elias Metschnikow for their ground-breaking work on immunity (https://www.nobelprize.org/prizes/medicine/1908/summary/).

Ehrlich's side-chain theory was rejected when experimental work led by Landsteiner demonstrated that upon injection of haptens (chemical groups which were covalently linked to the animal protein), antibodies specific for the haptenic group were produced and the production of the possible antibodies in an animal would be limited by the possible number of antigens (Landsteiner, 1945). With considerable amount of immunological data generated in addition to Landsteiners work, it was proven that any foreign substance that enters the body is an antigen and it was considered impossible that so many side chains of different specificities would be required by the body for absorbance of nutrients. The dismissal of side-chain theory led to the development of new hypothesis to understand the mechanism of antibody formation. Linus Paulings direct template theory assumed globulin as a single-polypeptide chain which folds around the antigen molecules resulting in formation of a high-specificity antibody molecule with stable configuration (Pauling, 1940). The specificity of the final antibody molecule is dependent on the complementarity with the antigen that serves as a template in the process of formation of an antibody molecule.

In 1955 Niels K. Jerne—a Danish immunologist—proposed the natural selection theory to explain the phenomenon of antibody synthesis and challenged the existing template theories. The natural selection theory suggested that globulins (antibodies) are continuously synthesized inside the body and few of the circulating globulins referred to as "natural antibodies" will have affinity toward an antigen. When an antigen enters the body, globulins will selectively bind to the antigens surface and few of them more or less fit well to the antigen resulting in an antigen−antibody complex which is engulfed by the phagocytic cell (Jerne, 1955). Upon entering the cell, the globulins dissociate from the antigen which plays no further part and antibodies stimulate the cell to produce more antibodies of the same kind. In addition to natural selection theory, Jerne's developed the plaque method and postulated the network theory to explain the antibody production process. The plaque method could help to identify and count the number of single antibody-producing cells among a mixture of cell population (Jerne & Nordin, 1963). Results from the experiments using mouse spleen cells and rabbit lymph node cells demonstrated that the number of obtained plaques on the Petri plate is directly proportional to the number of the lymphoid cells which eventually indicate the activity of the individual cells. Jerne's network theory was based on the idea that antibodies recognize and attach themselves not only to an antigen but also to other antibodies and interact with each other (antibodies) as part of a network and both antigen and antibodies can bind to the same site of the antibody (Jerne, 1974). In 1984 Jerne's was awarded the Nobel Prize jointly with George J.F. Kohler and César Milstein for their theories that explain the specificity in development and control of the immune system and the principle for production of mAbs (https://www.nobelprize.org/prizes/medicine/1984/summary/). The contributions of Kohler and Milstein will be discussed later in this section. Subsequently, in 1957 Frank Macfarlane Burnet introduced the clonal selection theory which was a modification of the Jerne's natural selection theory (Burnet, 1976). The clonal selection theory explains when an antigen enters the body, it binds to the lymphocytes cells that carry reactive sites corresponding to the antigenic determinants. This results in the activation of lymphocyte cells that further proliferate into a large number of clone cells (identical cells) to produce antibody against the foreign antigen.

Through the decades, a lot of significant discoveries were made and researchers continued to work on antibodies; however, the structure and its correlation to its function was still not clear. Due to the high-molecular weight of γ-globulin, it was not feasible to directly deduce its chemical structure and relate to its biological function. In order to tackle the structural problems of antibodies, R.R. Porter in 1959 attempted to use papain enzyme to cleave the γ-globulin molecule (attributed as antibody today) (Porter, 1959). The enzymatic treatment split the γ-globulin into three fragments with two of them constituting the antigen-binding sites, molecular weight around 50 kDa and demonstrating similar chemical and biological activity and no precipitating power. On the other hand, the third fragment is readily crystallizable with a molecular weight of around 80 kDa and no antibody activity. The two fragments with antigen-binding sites were Fab (fragment antigen binding), and the fragment that crystallizes was constant Fc (crystallizable fragment) domain. In the same year, Gerald M. Edelman intrigued by the structure of antibodies demonstrated that γ-globulin dissociate in the presence of denaturing agents indicating that γ-globulin subunits are linked by disulfide bonds though the possibility of the presence of other linkages was not eliminated

(Edelman, 1959). In 1972 both Porter and Edelman were jointly awarded the Nobel Prize in Medicine or Physiology for their discoveries about the chemical structure of antibodies (https://www.nobelprize.org/prizes/medicine/1972/summary/).

A decade later, Edelman established the entire amino acid sequence of a human γG1 Ig including the location of disulfide linkages and arrangement of LC and HC (Edelman et al., 1969). The data from the study showed that variable regions of LC and HC are homologous and similar in length. While the constant regions of the HC is composed of three homologous regions of similar length—C_H1, C_H2, and C_H3 which are also closely homologous to the constant region of the LC. However, the variable region of the chains is not homologous to the constant regions. Each variable and constant region is composed of one intrachain disulfide bond, and HC are linked via interchain disulfide bonds. The hinge region which lies in the middle of the HC (residue 221−233) contains two interchain disulfide bonds, rich in proline and cysteine residues and has no homology to any other portion of the antibody. The antibody structure will be discussed later in detail in Section 1.3.

mAbs were firstly isolated by the Argentinian biochemist César Milstein and his German postdoctoral fellow Georges Köhler back in the 1970s. Köhler while working at Milsteins laboratory was interested in finding the answer to the problem of finding the source of mAbs with predefined specificity. Along with the skills in studying myeloma in mice, Milstein's lab utilized a technique developed by Potter to establish a culture of rapidly proliferating tumor cells that produced antibodies or Igs (Potter, 1972). They successfully established and proved the technique to fuse the antibody-producing cells with myeloma cells. Kohler decided to fuse one myeloma cell bearing a selection marker with one mouse spleen cells, antibody-producing B-cells (Kohler & Milstein, 1975, 1976). A mutant myeloma cell line deficient in the enzyme hypoxanthine phosphoribosyl transferase was employed for the work. In order to identify the positive fused cells with hypoxanthine phosphoribosyl transferase activity, the fused cells were cultured in a medium containing hypoxanthine, aminopterine, and thymidine (HAT). Only the hybrid cells (also referred as hybridoma cells) survived and were selected as the enzyme contribution in the hybrid cells came from the healthy spleen cells (Fig. 1.1). These hybridoma cells survived as they inherited the immortality from myeloma cells and selective resistance from B-lymphocytes. For the first time, pure antibody specific to single antigen was produced in large quantities. This development of such an important method that demonstrates the fusion of antibody-producing cells from different origins

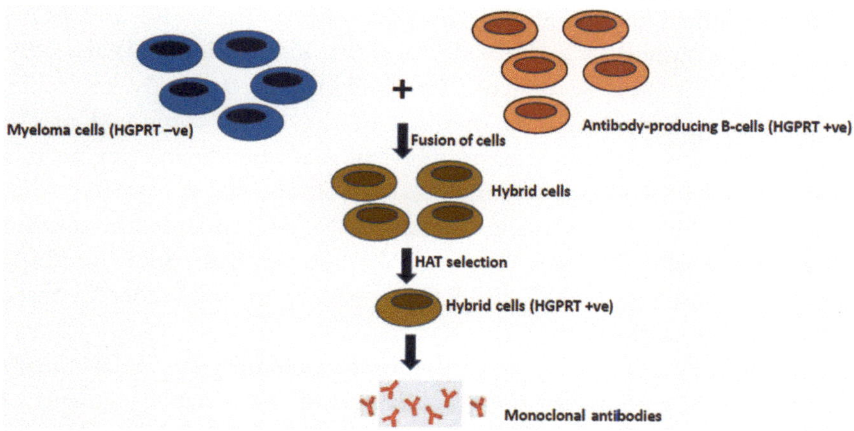

Figure 1.1 Monoclonal antibody production.

was never patented by Kohler and Milstein, allowing the use of this brilliant technology by biopharmaceutical industry and academicians for basic therapeutic research and bringing potential future therapeutic products for patients and also in diagnostic applications. For this breakthrough discovery, Milsten and Köhler along with Niels K. Jerne won a Nobel Prize in Medicine or Physiology in 1984 for their work on theories concerning the specificity in development and control of the immune system and the discovery of the principle for production of mAbs (https://www.nobelprize.org/prizes/medicine/1984/summary/).

1.3 Structure of monoclonal antibody

The basic structure of human mAb is a Y-shaped structure composed of two identical HC and two identical LC, of either kappa (κ) or Lambda (λ) type linked by disulfide bonds (Valentine & Green, 1967). The arms of the Y region of the antibody is the antigen-binding site and is commonly referred to as Fab (fragment, antigen-binding region). On the other hand, the tail of the Y region of the antibody is called the Fc (fragment, crystallizable) region which is always located on the C-terminal domain of the same antibody class.

The individual HC in the antibody has a molecular weight of 50,000 Daltons (Da), whereas the weight of each LC is 25,000 Daltons (Da). Individual LC is about 222 residues long with variable region (V_L) composed of 110 residues and constant region of the same length. The LC are covalently linked to the HC by disulfide bonds. Each HC and LC is made up of a number of globular domains that have variable (V) and constant (C) amino acid sequences. The bulk of the antibody molecule, three-fourths of the HC, and one-half of the LC is composed of the highly conserved constant regions. The HC has one variable region followed by several constant regions (C_H1, C_H2, and C_H3). On the other hand, the LC has one variable domain (V_L) and one constant domain (C_L). The variable domains (V_H and V_L) and constant domains (C_H1, C_H2, C_H3, and C_L) of HC and LC are similar to each other in three dimensional conformation.

The variable region is part of the antigen-binding site and antibody specificity is determined by the combination of the variable segments of the HC and LC. The variable regions of HC and LC fold together to form the hypervariable loops that form the antigen-binding site or previously mentioned as "paratope." The variable region of the HC is directly connected to the HC C_H1 domain. The C_H1 and C_H2 domain of the HC constitutes one of the most fascinating regions in the antibody structure, also called as the "Hinge Region" (or "Interdomain Region") (Adlersberg, 1976). The hinge region is an amino acid stretch rich in cysteine and proline residues located in the central region of the HC that links the two HC by two disulfide bonds in an IgG molecule. The hinge region has no similarity to any other region in the antibody structure and is an extremely variable amino acid sequence from one IgG class to another. The hinge region is the primary mediator to the flexibility of the antibody—the larger the hinge region, the more flexible is the antibody, thus allowing better antigen–antibody interactions and good effector activity (Roux, 1997; Vidarsson, Dekkers, & Rispens, 2014). The upper hinge region connects the Fab arms to the middle hinge region and is responsible for the Fab–Fab segmental flexibility with length of the upper hinge region significantly influencing the Fab segmental flexibility (Roux, 1997; Schumaker, 1991). On the contrary, the middle hinge region which is rich is in cysteine and proline residues typically with the sequence –Cys–Pro–Pro–Cys– is relatively inflexible and serve as spacer that separates Fab from Fc arm in length-dependent manner (Matsunaga, 1991). Lastly, the lower hinge region which is less flexible compared to the upper region is responsible to connect the Fc to middle hinge region and contributes to Fab-Fc flexibility and Fc "tail wagging" movement required for Fc receptor binding. Compared to the rigid middle hinge region which is rich in cysteine and proline residues, the upper and lower regions are rich in highly flexible peptide segments. Roux et al. has defined different modes of flexibility in the hinge region-folding, rotation along the long axis, conical wagging, and translation (in- and out- motion) (Roux, 1997). Additionally, results from another study demonstrate that the hinge region particularly the upper hinge region undergoes an extensive degree of internal motion as obtained using high-resolution nuclear magnetic resonance analysis (Kim, 1994). The data from the same study have also shown that longitudinal interactions between C_H2 domain and C_H3 domain remain unaffected by the deletion of the Fab segment, cleavage of interchain disulfide bonds, and deletion of the most of the hinge region which is in alignment with the previously reported work about the protein A binding to IgG1 myeloma proteins DoB and Lec (Arata, 1980; Klein et al., 1981).

The flexibility of the hinge region makes it the weakest link in the antibody structure and extremely vulnerable to different mechanisms of antibody degradation pathways such as proteolytic enzymes attack or non-enzymatic proteolytic cleavage such as direct peptide bond hydrolysis or metal ion catalyzed cleavage leading to IgG fragmentation that might affect the product integrity and protein function (Brezski et al., 2009; Cordoba, Shyong, Breen, & Harris, 2005; Gearing, 2002; Xiang, 2007). A selective cleavage study on human IgG demonstrated the two of the matrix metalloproteinases (MMP), matrilysin, and stromelysin-1 cleave Ig at a conserved site located between the CH_2 domain and hinge region of the constant domain of the HC resulting in generation of a single-chain Fc-like fragment of molecular weight 32 kDa (Gearing, 2002). The amino acid sequence analysis data indicated that matrilysin (referred to as MMP-7) cleaves between the two leucine (L) residues, whereas the stromelysin-1 (referred to as MMP-3) cleaves between proline (P) and glutamic acid (E) residues in all the subclasses of the IgG. Brezski et al. have shown that effector functions of the IgG particularly IgG1 may be compromised due to the single proteolytic cleavage in the lower hinge region of the antibody (Brezski et al., 2009). Compared to the intact IgG, the single cleaved IgG demonstrated more than 20-folds losses of the relative binding affinities toward different FcγRs that mediates antibody-dependent cellular cytotoxicity (ADCC) and complement-dependent cytotoxicity (CDC) activities. Not limited to proteolytic cleavage, studies indicated that hinge region peptide bond cleavage can also be accelerated by the acidic and basic conditions of the local environment with no change in cleavage rate in the presence of protease inhibitors and host cell proteins

(HCPs) (Cordoba et al., 2005; Xiang, 2007). The mass spectrometry data revealed major cleavage sites between amino acids Asp (D) and Lys (K), and His (H) and Thr (T) in the hinge region of the intact mAb and size exclusion chromatography results showed that hinge region amino acid sequence −Asp−Lys−Thr−His−Thr− (−DKTHT−) was prone to acid- and base-catalyzed cleavage with no impact of protease inhibitors and EDTA (Cordoba et al., 2005). The cleavage sites with amino acids Asp (D) and His (H) in the protein sequence were susceptible to acid- and base-catalyzed hydrolysis. The novel cupric ion (Cu^{++}) catalyzed peptide bond cleavage in human IgG1 Campath 1H has been reported (Smith et al., 1996). This metal catalyzed chemical cleavage increased at alkaline pH (pH 8.0), higher temperature (62°C) and with an increase in cupric chloride concentration (10 mM $CuCl_2$). However, this copper-induced cleavage could be inhibited by addition of EDTA. The results from partial N-terminal analysis of the study indicated that the specific cleavage site was $Lys_{226}-Thr_{227}$ bond (trypsin cleavage site) in the hinge region sequence −DKTHT−. To conclude, the findings from this study demonstrated that antibodies or proteins should not be stored or exposed to small amounts of copper.

Papain and Pepsin are two of the most common enzymes that cleave in or close to the hinge region and are commonly employed for antibody based fragmentation applications for efficient tissue penetration, avoidance of nonspecific antigen interactions, and simpler systems for structural studies. Papain cleaves at neutral pH above the disulfide bonds in the hinge region, generating two identical Fab fragments each containing one LC disulfide bonded to the HC (F_d) containing the variable region (V_H) and C_H1 region (Fig. 1.2). The remaining portion is the Fc region constituting the C_H2 and C_H3 domains. Pepsin is an another common enzyme that cleaves at acidic pH below the disulfide bonds that connect the two HC in the C_H2 region. The resulting fragments consist of two $F(ab')_2$ fragments linked with disulfide bonds (Fig. 1.2). However, no intact Fc fragment is produced as it is largely degraded by the pepsin with no biological function.

The hydrophobic face of the C_H2 domain on both HC in the Fc portion is partially covered with a N-glycan moiety at Asn297 position that contains the consensus sequence −Asn−X−Ser/Thr where X could be any amino acid except proline. This attachment of the oligosaccharide unit to the N-terminal of asparagine residue is commonly termed as "N-glycosylation" and accounts for 2%−3% of the antibody's molecular

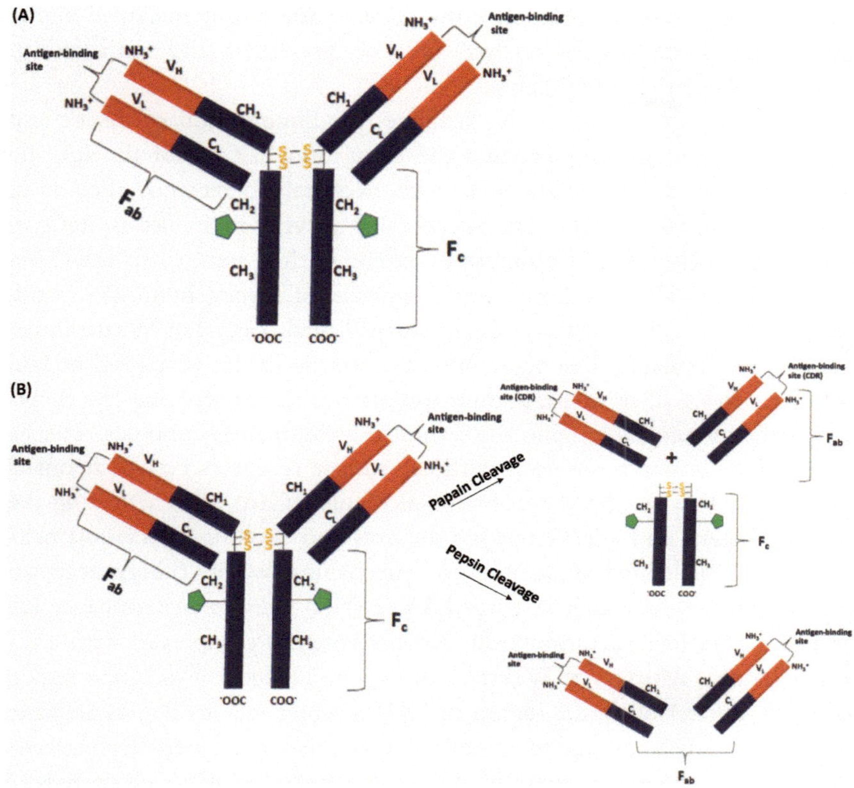

Figure 1.2 (A) Structure of IgG. (B) IgG obtained after papain and pepsin digestion. *IgG*, immunoglobulin G.

weight. In contrast to N-glycosylation, O-linked glycosylation is the sequential addition of individual monosaccharide unit, to the oxygen (−O−) atom in hydroxyl (−OH) group of the sides chains of serine (Ser, S) and threonine (Thr, T) amino acid residues of the protein (Kaur, 2021; Van den Steen et al., 1998). Glycosylation is a posttranslational modification (PTM) that plays an essential role in antibody effector functions, interaction of Fc domain with Fc receptors (FcγRs) and complement molecule (C1q), antibody stability, conformation, and pharmacokinetics (Alessandri et al., 2012; Lis & Sharon, 1993; Thomann, Reckermann, Reusch, Prasser, & Tejada, 2016; Yu et al., 2014). The impact of each of the glycan forms (fucosylation, galactosylation, terminal sialic acid, and high mannose) on the antibody effector function and its biological behavior will be discussed in detail in Chapter 3: Glycosylation Analysis. In addition, the factors that influence the glycan pattern and analytical methods employed for glycan analysis will be presented and discussed in Chapter 3: Glycosylation Analysis.

The C_H3 domain of the HC facilitates the dimerization of the Fc and along with C_H2 domain, increases the circulation half-life of the IgG by contributing to the Fc interactions with neonatal Fc receptor (FcRn) in acidic environment of the endosomes and recycling circulating IgG in serum at physiological pH resulting in extending IgG serum half-life (Ying et al., 2013). Each C_H3 domain is composed of several hydrogen bonds between polar side chains and extensive lateral van der Waals forces between the domains. The two constant domains of the Fc region of IgG HC, C_H2 and C_H3 is of paramount importance in determining the therapeutic efficacy and prolonging the serum half-life of the antibodies. The Fc region of the antibody interacts with cell surface receptors FcγRs, neonatal Fc receptor (FcRn), and complement molecules (C1q), thus activating the immune system and elicit effector functions and prolong circulation half-life. ADCC is defined as an immune mechanism where the Fc receptor bearing effector cells such as natural killer (NK) cells or macrophages kill the target cells (e.g., a tumor cell) that has antigens complexed with antibody (Roman, Murray, & Weiner, 2014). The interaction of hinge region and C_H2 domain of the Fc region of IgG1 antibodies with FcγRs activates the effector cells resulting in release of molecules contained in cytotoxic granules such as perforin, granulysin, and granzymes which leads to lysis of the target cells. The ADCC mechanism is highly dependent on the binding affinities of the Fab region of the antibody to the target molecules and binding affinities of the Fc region of the antibody to FcγRs. The Fc receptors (FcγRs) that are commonly expressed on the effector cells to mediate the ADCC activity includes FcγRI (CD64), FcγRII (CD32), and FcγRIIIa (CD16) (Roman et al., 2014). Of these, NK cells predominantly expresses FcγRIIIa which bind to the Fc portion of the Ig. Studies have demonstrated ADCC as the key mechanism of action (MoA) for killing the cancer cells in Trastuzumab antibody (Musolino et al., 2008). In addition, CDC is another major effector function that involves the binding of the Fc region of the antibody to the complement proteins (C1q), thus activating the complement cascade proteins and resulting in the formation of the membrane attack complex and eventually target cells lysis (Bindon, Hale, Brüggemann, & Waldmann, 1988; Sarma & Ward, 2011).

In recent years, extensive efforts have been made to modulate the effector functions of the mAbs to improve their drug action and enhance their therapeutic efficacy (Natsume, Niwa, & Satoh, 2009; Pereira, Chan, Lin, & Song, 2018; Ricklin & Lambris, 2007). In order to enhance the ADCC effector function, studies have been reported where antibodies have been modified by glycoengineering of the Fc N-glycans or by engineering the Fc domain directly. The antibodies produced by Chinese hamster ovary (CHO) cells displays glycan heterogeneity with G0F being the most abundant N-glycan species. Overexpression of galacosyltransferses to a certain limit in CHO cells could be useful in increasing the G2F percentage and thus enhancing the ADCC activity (Pereira et al., 2018). However, the data from another study indicates the addition of galactose to afucosylated antibodies does not increase the ADCC effect indicating afucosylation the key mediator of the ADCC (Shinkawa et al., 2003). The researchers in this work produced two antibodies antihuman interleukin 5 receptor (hIL-5R) humanized immunoglobulin G1 (IgG1) and an anti-CD20 chimeric IgG1 from rat hybridoma YB2/0 cells with low fucose content (less expression of fucosyltransferse, FUT-8 mRNA) and CHO cells that have the characteristic of having high fucose content. The antibodies generated from YB2/0 showed 50-folds higher ADCC than those produced from CHO cells. Shields et al. engineered the Fc domain of the antibody by mapping the residues in IgG1 involved in the binding to human FcγRs and FcRn and replacing the solvent exposed residues in C_H2 and C_H3 regions with alanine residues which is near to the IgG1: FcγRIII interface (Shields et al., 2001). The substitution with alanine showed affinities of the residues vary for different Fc receptors with improved binding to one receptor and reduced binding to the other type. The selected variants S298A, K334A, E333A, and T256A demonstrated improved binding affinities for FcγRIIIa, and thus enhancement in the ADCC activity when NK or peripheral blood monocyte cells were used. In addition to modulating the ADCC function, several approaches have been utilized to improve the CDC effector function by enhancing the binding interactions between the C1q and Fc domain of the antibody. Idusogie and colleagues employed site-directed mutagenesis to improve the CDC activity of the anti-CD20 antibody Rituximab (Idusogie et al., 2001). The substitution of the two sites, K326 and E333, located in the C_H2 domain of the Rituximab improves the CDC activity revealing that point mutations could have tremendous potential in developing therapeutic mAbs. In another study, engineering of the hinge region in a human IgG1 antibody by site-directed mutagenesis resulted in modulating the hinge length, its flexibility, and antibody's effector functions (Dall'Acqua, 2006).

The data from the study showed that C1q activity of the antibody is highly influenced by the rigidity in the middle region and is unaffected by the flexibility of the upper hinge region. Additionally, deletion of two amino acids in the middle hinge region significantly impacts the binding of the Fc region to C1q but is unaffected with deletion in the upper region. Thus engineering the hinge region could be an attractive target to tune the effector function in human IgG1 and develop therapeutic antibodies with enhanced CDC activity.

The Fab region is composed of two variable regions (V_L and V_H) and two constant regions (C_L and C_H). Within each variable region are six hypervariable loops, three in LC and three in HC, supported by conserved four framework regions (FR) of β-sheets (Davies & Chacko, 1993). The hypervariable regions are a stretch of amino acids that are commonly known as CDRs and are always located on the N-terminal region of the antibody of the same antibody class. The CDRs are part of the paratope region of the antibody that are involved in a myriad of noncovalent interactions with the epitope region of the antigen and are responsible for antigen recognition and specificity and affinity of the antibody. The nature of noncovalent bonds that favor antigen—antibody complex formation includes electrostatic interactions (columbic forces), van der Waals forces, hydrophobic interactions and short-range hydrogen bonds (van Oss et al., 1986). The antigen combining site is formed by six CDRs appearing as loops from one end of the β-sheet, three from HC (V_H), and three from LC (V_L). Of the six CDRs together involved in antigen binding, each of them have their own amino acid composition and length and also significantly differ in the number of antigens they bind and amino acid residues on the antigen surface. This could be the result of evolutionary response allowing antibody to recognize any patch of the antigen surface (Kunik, Good, & Chaudhury, 2013).

The FR are non-CDR parts of the variable region of the antibody which have less variability compared to CDRs (or hypervariable) sequences and are more conserved. The FR residues may or may not be the part of the antigen-binding process. The FR residues that are in direct contact with the antigen (or part of the antigen-binding region) are generally in close proximity to the CDRs or within the boundaries of CDRs. On the other hand, the FR residues that affect the Ag binding but are not in direct contact with the Ag may or may not be in the spatial proximity to the CDRs with the former providing the structural support to the CDR loops enabling them to adopt the right conformation and orientation required for Ag binding (Kettleborough, Saldanha, Heath, Morrison, & Bendig, 1991; Sedrak, Hsu, & Mohan, 2003; Xiang, Sha, Jia, Prasad, & Delbaere, 1995). With some amino acids more likely to form noncovalent interactions with antibody to stabilize the protein–protein interaction, a detailed analysis of frequency of occurrence of amino acid residues in CDRs that are in direct contact with the antigenic determinant (or epitope) and amino acids that have conformational influence on contacting residues particularly in the FR was attempted (Padlan, 1990). The data reveals the presence of asparagine Asn (N), histidine (H), and tyrosine (Y) residues is more prevalent in CDRs, whereas the cysteine is more likely to be present in FR. Furthermore, the nonpolar aliphatic amino acids such as alanine (A), valine (V), isoleucine (I), and Leucine (L) are more likely to occur in FR compared to CDRs. The aromatic amino acid residues such as phenyalanine (F), tyrosine (Y), and tryptophan (W) are more exposed to solvent in CDRs than when buried in the FRs and thus contribute to the antigen-binding surface.

Considering the crucial significance and implications of CDRs for Ag recognition and binding, extensive efforts have been made for the identification of CDRs, their boundaries, and composition that could assist to identify paratopes on the antibody surface. In addition, research work concentrated on the sequential numbering of the residues has been conducted to determine the residues within the protein structure and the sequence. Using the available structural data on the LC and HC, Wu and Kabat were the first to attempt to locate the precise positions of the CDRs in the variable regions and its unique features in Igs (Wu & Kabat, 1970). Analysis of the residues in the variable regions shows the presence of glycine (G) of the variable region significantly confers the flexibility to antibodies due to lack of the side chain thus allowing motion of the antibody with minimal steric hindrance. The data from the study also defined three regions with high variability in the LC and HC—residues 24–34, 50–56, and 89–97. Of these, positions 24–34 and 89–97 constitute the CDRs of the HC and LC and form the antigen-binding site. Chotia et al. identified the set of amino acid residues that are responsible for the conformation of the CDR and FRs in Igs and has shown that the presence of certain key conserved residues in the CDR and loops of the FRs of different Igs result in particular conformation (Chothia & Lesk, 1987; Chothia et al., 1989). This indicates that CDRs (or hypervariable regions) that have the same conformations in different Igs have the same or very similar residues at these sites. Additionally, this structural model can be implied to determine the main chain conformations in the hypervariable regions in unknown sequences by inspecting the similar residues in the known sequences. Lefranc and colleagues introduced a highly consolidated and comprehensive database—IMGT, the international ImMunoGeneTics database for Igs and T-cell receptors (TcR) variable domains (Lefranc et al., 1999, 2003). Using the IMGT numbering scheme, amino acid residues located in CDRs and FRs in different sequences can be compared without the alignment and length and limits of the CDRs and FRs can be defined thus allowing the characterization of the variable regions of the antibodies. In 2001 another numbering system referred to as AHo method with special focus on the variable domains of the Igs and TcR was devised. Unlike the IMGT scheme where the insertions and deletions grow unidirectionally, the AHo numbering scheme places the same symmetrically at key positions (Honegger & Plückthun, 2001). Furthermore, the length variations in the CDR 1 and CDR 2 is represented by one or two gaps in AHo compared to one gap in IMGT. Overall, the insights and information generated from variable domain analysis in Igs could be extremely valuable in construction and sequence analysis of engineered antibody fragments with specificity and in vitro evolution of antibodies.

Currently, in the biopharmaceutical industry, one of the most common methods employed for identification of CDRs and determining the CDR sequence is referred to as "Peptide Mapping." Peptide Mapping is a key analytical method used to map the protein sequence by mass spectrometry via enzymatic digestion of protein and is used as proof of identity for lot release testing by confirming the primary structure (amino acid sequence) of the protein (or mAb). Peptide mapping is also used to identify the modifications termed as "Posttranslational Modifications (PTMs)" in amino acid residues in the CDRs and other domains of the antibody. The protein primary sequence identity and PTMs in particular are considered one of the critical quality attributes that might affect the safety, efficacy, and overall therapeutic behavior of the drug. This topic will be discussed in detail in Chapter 4: Pimary Structure and Peptide Mapping.

1.4 Therapeutic applications of monoclonal antibodies

Over the last four decades since their discovery, mAbs have made striking advances from essential research tools to life-saving therapeutics. To date, the regulatory agencies across the globe have approved all types of mAbs—murine, chimeras, and humans. In the early days of mAbs production, animals such as murine (typically mouse or rat) is injected with the target antigen and B-cells producing the therapeutic mAb against the target antigen were extracted by harvesting the spleen cells of the animal. Muromonab-CD3 (trade name Orthoclone OKT3), the first US Food and Drug Administration (FDA)-approved mAb to become available for therapy in humans was created using this method. OKT3 was approved in 1985 to reduce the chances of rejection from kidney transplant in patients. The approval of OKT3 was considered one of the landmarks not only in the field of clinical transplantation of solid organs but also in the field of therapeutics. OKT3 specifically targets CD3 that is found only on mature T-cells and medullary thymocytes (Norman, 1995). OKT3 binds and subsequently removes the CD3-TCR complex from the T-cell surface and thus prevents the presentation of T-cells to allograft antigens that contributes to organ rejection. However, an adverse side effect of OKT3 infusion is T-cell activation that leads to release of cytokines and thus causes cytokine release syndrome. Murine mAbs including OKT3 posed a serious risk of immunogenicity as they were recognized as foreign material by human body resulting in production of human antimouse antibody (HAMA), thereby limiting their use in clinical trials and henceforth, their development as therapeutic molecule (Schroff, 1985; Shawler, 1985). To overcome immunogenicity risks, novel approaches and technologies were developed to engineer antibodies and make them more human like or completely human. Although initial executions were not satisfactory, attempts were made to increase biological efficacy and decrease the immunogenicity by either generating chimeric or humanized mAbs (derived from human and mouse DNA) through genetic engineering and expressing in mammalian, plant, or insect cells. Another approach involved production of human mAbs (fully human) is through phage display technology that will be discussed later in this section.

Morrison et al. demonstrated the creation of chimeric human antibody molecules by ligating the variable region genes from mouse antibody-producing myeloma cell line with known antigen binding specificity to human Ig (IgG1 or IgG2) constant regions genes by recombinant DNA technology (Morrison, 1984). The HC variable region from mouse myeloma was joined to HC constant region from human Ig and LC variable region exon from same myeloma was joined to human κ LC gene. The chimeric gene construct was transfected into mouse myeloma cell line resulting in the production of novel human antibody molecules that are largely human with antigen binding specificities from mouse. In the same year, another study showed the production of chimeric antibody by joining DNA segments encoding for the mouse variable region specific for trinitrophenol (TNP) to segments encoding human μ and κ constant regions (Boulianne, 1984). The generated Ig was chimeric IgM specific for TNP and can activate the complement. Chimeric antibodies were thought to be a good substitute for murine mAbs as they could decrease or eliminate the immunogenicity risk (or HAMA response) by making the antibody product invisible to patient's immune system. Additionally, chimeric antibody allows joining of same variable region (V) to different human antibody molecules constant (C) regions thus exploring the use of C region to trigger desired biological effector functions. In a similar manner, the V region could be modified to alter their interaction with antigens and generate antibodies with different antigen binding specificities.

In 1997 Rituximab (Trade name: Mabthera/Rituxan, Genentech) was the first chimeric mAb (human constant region and mouse variable region) produced in CHO cells that was approved by US FDA for the treatment of the relapsed or nonrefractory indolent non-Hodgkin's lymphoma (NHL). Rituximab binds to the CD20 antigen expressed on the surface of 95% of B-cell lymphoma cells and normal B-cells. The binding of Rituximab to CD20 triggers cell death via multiple mechanisms which include indirect methods such as ADCC, CDC, or directly binding to CD20 and inhibiting the cell proliferation or inducing apoptosis (Cerny, Borisch, Introna, Johnson, & Rose, 2002). However, the exact MoA of Rituximab is not well understood at in vivo level including the interdependency of different effector functions and direct inhibition. Since its first approval in 1997, the therapeutic application of Rituximab as a monotherapy or in combination with traditional chemotherapeutics has extended to multiple indications including different types of cancer and autoimmune disorders (Plosker, 2003; Salles et al., 2017; Schmidt, 2006). Due to its widespread success in treatment of a wide range of diseases, Rituximab was one of the top 15 selling drugs in 2018 and is listed as the World Health Organization (WHO) list of essential medicines that are considered safe, efficacious, and coeffective medicines needed for basic healthcare system (Philippidis, 2019). With loss of data exclusivity and patent protection expiry to innovator Rituximab in Europe in 2013 and in the United States in 2016 has led to the development of safe, efficacious, and well-characterized biosimilars to Rituximab by different companies globally (WHO, 2015; Generics and Biosimilars Initiative (GaBI) Online, 2018b).

The superiority of chimeric antibodies over the murine mAbs in terms of therapeutic efficacy drove the researchers to develop antibodies that constitutes a greater percentage of human sequence or completely derived from a human source. Shortly after the production of chimeric mAbs was demonstrated. Jones et al. attempted to graft the CDRs of the HC variable region from mouse antibody which was specific for NP-cap (4-hydroxy 3-nitrophenacetyl caproic acid, $K_d = 1.2\ \mu M$) with corresponding CDRs of a human myeloma protein (Jones, Dear, Foote, Neuberger, & Winter, 1986). The affinity of the CDRs of the HC for a related molecule NIP-cap (3-iodo-4-hydroxy-5-nitrophenyl-acetyl caproic acid) was $0.02\ \mu M$. The new antibody produced from the gene construct formed by linking the CDRs of human myeloma with mouse antibody LC, demonstrated the binding affinity of $K_d = 1.9\ \mu M$ for NIP and $K_d = 0.07\ \mu M$ for NIP. With no significant change in the binding affinities and specificity on CDR replacement, this grafting method can be used for constructing human antibody from an appropriate mouse antibody and henceforth applied for the production of humanized antibodies. The humanized mAb construct constitutes the CDRs (or antigen-binding regions) derived from mouse, and rest of the variable region and constant region derived from a human source. On the other hand, the chimeric mAbs are produced from the variable regions that is completely derived from a mouse and constant regions derived from a human source.

Human mAbs are another class of mAbs derived wholly from a human source. Unlike the humanized antibodies where the production always starts with immunization of the wild type mice with intact mice Ig locus, the production of human antibodies starts with immunization of animals using transgenic mice or phage display. In the case of human antibodies, a transgenic mouse which has been genetically engineered with human Ig locus is injected with specific target protein and specific antibodies generated against the target are identified and CHO cell lines are used to produce the specific antibodies (Mallbris et al., 2016). Another method of generating human mAbs is using phage display (Schirrmann, Meyer, Schütte, Frenzel, & Hust, 2011). The phage display method involves the isolation of mRNA from B-lymphocytes of the human blood and converting into cDNA using polymerase chain reaction to amplify V_H and V_L segments. Prior to infecting *E. coli*, the generated variable segments are cloned in a bacteriophage vector [usually as single-chain fragment variable (scFv) or fragment antigen binding (Fabs)] to generate a library with large counts of cells. The variable fragments (V_H and V_L) that are secreted as part of the bacteriophage coat from the *E. coli* can be selected for a specific antigen and are used to reinoculate the *E. coli* with the bacteriophage. Cells containing the plasmid can then be isolated and sequenced. Compared to the immunization method, the phage display is advantageous as no immunization is required and the entire selection process is in vitro. Additionally, the same library can be used to generate new antibodies. The phage display that is commonly used nowadays was first described by George Smith in 1985 when he displayed that insertion of foreign sequence in between the two domains of the phage filamentous gene encoding for minor coat protein (Gene III) results in the formation of a fusion protein which when inserted into a virion generates a "fusion phage" which is 1000-fold more enriched than ordinary phage by affinity for antibody directed against the foreign sequence (Smith, 1985). This technique can be used for cloning of a gene from random sequence when the antibody against the product is available. In 1990 similar promising studies used phage display technology where variable (V) regions of antibody were displayed on the surface of the bacteriophage. The phages were selected by binding specifically to an antigen and soluble antibody fragment could be secreted by an infected bacteria (McCafferty, Griffiths, Winter, & Chiswell, 1990; Winter, Griffiths, Hawkins, & Hoogenboom, 1994). George P Smith and Sir Gregory P. Winter were jointly awarded the Nobel Prize in Chemistry in 2018 along with Frances Arnold for their work on phage display technique (https://www.nobelprize.org/prizes/chemistry/2018/press-release/; Winter, et al., 1994). George P Smith was awarded the Nobel Prize for developing the phage display method, whereas Sir Gregory P. Winter was awarded to use the phage display for the production of antibodies for therapeutic uses. The last two decades have witnessed successful application of phage display technology for development of therapeutic mAbs. In 2002 Adalimumab (Trade name: Humira) was the first therapeutic fully human mAb approved by US FDA that was derived from phage display technology (Tayar, Lopez-Olivo, & Suarez-Almazor, 2013). Since then, phage display has produced antibodies that can neutralize toxins, counteract autoimmune diseases, and cure metastatic cancer. Adalimumab will be described in more detail later in this section.

Apart from design of an antibody construct and method of production, the humanized and human mAbs can be distinguished from the chimeric mAbs from their nomenclature. The International Nonproprietary Names (INN) of mAbs is assigned by WHO and is supposed to have a prefix, substem A, substem B, and suffix (General policies for monoclonal antibodies, 2009; Tayar, Lopez-Olivo, & Suarez-Almazor, 2013). The common stem used for all mAbs is "mab," placed as a suffix and prefix can be random but eventually must contribute to a distinctive name. The substem B which precedes the stem denotes the animal/species from which antibody is obtained and substem A that precedes the substem B indicate the target molecule—molecule, cell, or organ. The chimeric antibody name contain a −xi− substem-B such asRitu*xi*mab, Infli*xi*mab, and Cetu*xi*mab, whereas the humanized antibody name contain a −zu− substem B, for example, Trastu*zu*mab, Bevaci*zu*mab, and Omali*zu*mab, and human antibody contain a −u− substem B, for example, Adalim*u*mab, Golim*u*mab, and Ipilim*u*mab.

In 1998 Trastuzumab (Trade name: Herceptin, Genentech), a humanized IgG1 kappa mAb was approved for the treatment of metastatic HER2 overexpressing breast cancer. Trastuzumab consists of two antigen-binding domains that binds to the extracellular juxtamembrane domain of the human epidermal growth factor receptor (HER2 or HER2/Neu or ErbB-2) that is overexpressed in some breast cancer cells and decreases the signaling through several possible mechanisms such as downregulation of HER2 disrupts HER2 dimerization, cell cycle arrest in the G1 phase with concomitant reduction in cell proliferation, release of extracellular domain due to the proteolytic cleavage of HER2/Neu, and recruitment of immune effector cells responsible for ADCC function (Hudis, 2007). HER2 is an important biological target for breast cancer and accurate assessment of the expression levels of HER2 in an essential parameter for Trastuzumab targeted therapy treatment in breast cancer patients. Trastuzumab as a single agent or in combination with chemotherapy has brought significant advantages to breast cancer patients and has had an enormous impact in the management of HER2 positive metastatic breast cancer. With international demand for oncology medications, Trastuzumab has reached global sales of $7 billion revolutionizing the breast cancer treatment and has been declared one of the blockbuster drugs in the global market. Additionally, Trastuzumab was one of the top 15 selling drugs in 2018 and is listed as the WHO list of essential medicines that are considered safe, efficacious, and coeffective medicines needed for basic healthcare system (Philippidis, 2019). The high cost and loss of exclusivity of Trastuzumab in 2019 led to development of several biosimilar versions of Trastuzumab in the United States, Europe, and Asian markets (Barbier et al., 2019). The creation of biosimilars may assist to make medicines (or drug therapies) affordable and accessible to the patients and reduce pressure on healthcare budgets.

Bevacizumab (Trade name: Avastin, Genentech) is the humanized antiangiogenic mAb that was approved in 2005 by US FDA for the first line treatment of metastatic colorectal cancer in combination with chemotherapy (Ferrara, 2004; Presta, 1997). Bevacizumab binds to vascular endothelial growth factor (VEGF-A) that plays a key role in the regulation of normal and pathological angiogenesis. The binding of Bevacizumab to VEGF-A blocks the binding of the VEGF protein to its cellular receptors (VEGFR-1 and VEGFR-2) on the surface of endothelial cells, resulting in inhibition of endothelial cell mitogenic activity, reduction of microvascular growth of tumor blood vessels and thus limiting the blood supply to tumor tissues and leads to apoptosis of the tumor endothelial cells (Shih, 2006). Since its approval in February 2005, Bevacizumab treatment has been extended for additional malignancies including nonsmall cell lung carcinoma, renal cell carcinoma, glioblastoma cancer, ovarian cancer, and cervical cancer, resulting in regulatory approvals in the United States and in other countries (Ferrara, 2016). Bevacizumab is also used as an off label drug for ophthalmic diseases. More than 2 million patients have been treated with bevacizumab, and today, this drug ranks among the most widely used therapeutics in oncology. Along with Trastuzumab, Bevacizumab was also one of the top 15 selling drugs in 2018 and is listed as the WHO list of essential medicines that are considered safe, efficacious, and coeffective medicines needed for basic healthcare system (Philippidis, 2019). With the expiry of the innovator molecule, many Bevacizumab biosimilars versions for different indications are commercialized in the market to improve the healthcare access to patients by increasing options of medications and potentially lower costs (Ferrara, 2004).

Adalimumab (Trade name: Humira, Abbvie) is the first fully human mAb (IgG1: κ constant regions) approved in 2004 for treatment of rheumatoid arthritis (RA) (Generics and Biosimilars Initiative (GaBI) Online, 2019). It is produced by recombinant DNA technology in a mammalian expression system and has a molecular weight of 144 kDa. Adalimumab is a tumor necrosis factor (TNF-α) inhibitor which is one of the inflammatory cytokines induced by RA and is found in very high concentration in the synovial joints. TNF-α is naturally occurring cytokine which at higher concentrations lead to inflammation and organ damage. Adalimumab MoA involves specifically binding to TNF-α and blocking the interaction between TNF-α and its two receptors (p55 and p75) and thus neutralizing the activity of TNF-α and reducing inflammation in joints (Keystone & Haraoui, 2004). In vitro studies have indicated that Adalimumab lyse cells expressing high levels of surface membrane TNF-α in the presence of the complement. Adalimumab has been FDA approved for several indications including RA, Juvenile Idiopathic Arthritis, Psoriatic Arthritis, Ankylosing Spondylitis, Adult Crohn's Disease, Pediatric Crohn's Disease, Ulcerative Colitis, and Plaque Psoriasis (Keystone & Haraoui, 2004). Adalimumab is one of the blockbuster drugs and considered one of the world's most successful drugs with global sales peaking at $19.9 billion in 2018 (Philippidis, 2019). However, Adalimumab patent expiry in 2016 has opened doors for affordable biosimilars with few of the in development and clinical pipeline and handful of them launched in Indian market (Adalimumab: Highlights of Prescribing Information (FDA). 2018). Additionally, this drug is listed as the WHO list of essential medicines that are considered safe, efficacious, and coeffective medicines needed for basic healthcare system (Philippidis, 2019).

In addition to chimeric, humanized, and human antibodies, there are another class of novel recombinant therapeutic proteins called as "fusion proteins" that are constructed through joining of two genes that originally coded for separate proteins. By genetically fusing two or more proteins together, the fusion product would contain the distinct functions of each of the component moieties. Etanercept (Trade name Enbrel or Benepali, Immunex/Amgen) is a fully human, recombinant dimeric fusion protein consisting of two copies of extracellular ligand-binding portions of the human 75 kDa soluble TNF-α receptor linked to Fc fragment of IgG1 (Strober, 2005). The Fc fragment constitutes C_H2 and C_H3 domain but not the C_H1 domain and assist in prolonging the half-life of molecule in the bloodstream thereby resulting in long-lasting biological effect compared to naturally occurring soluble TNF-α receptor only. Etanercept was the first TNF-α inhibitor approved by US FDA for treatment of psoriatic arthritis and plaque psoriasis in 1998. Etanercept is a TNF-α receptor blocker that exhibits its MoA by acting as a decoy receptor and inhibits the binding of the soluble TNF-α to TNF-α receptor by binding itself to TNF-α (Strober, 2005). The dimeric structure of the TNF-α allows Etanercept to bind to two TNF-α molecules simultaneously and enhances its binding affinity by 50- to 1000-folds higher than the naturally occurring soluble receptor (Molta, 2007). Etancercept is one of the top selling and lucrative drugs in the pharmaceutical market and secured global sales of around 7.1 billion in 2018 (Strober, 2005). Being one of the costliest drugs in the healthcare system and patient expiry in the Europe in August 2015, several biosimilars have been launched in different markets for treatment of RA (Philippidis, 2019).

1.5 Conclusions

It is noteworthy that the range of therapeutic mAbs approved by regulatory authorities have dramatically increased with each passing year. The advent and use of mAbs has strongly influenced the medical care of patients and revolutionized the treatment of diseases such as cancer, respiratory, cardiovascular, and autoimmune diseases and infections. With more than 80 mAbs approved and multitude of them in different stages of the clinical trials and development pipeline, they have finally come of age as therapeutics offering promising new treatments to patients worldwide and leading to significant changes in healthcare arena. Although mAbs present an attractive option for development of new therapies and molecular drug targets against a wide variety of diseases, high cost of the approved mAbs is still a challenge to be overcome particularly in developing nations. Efforts should be made to reduce the cost of existing drug manufacturing and purification processes through use of improved and advanced technologies. Additionally, private companies and academia should collaborate to exchange information, ideas, and develop therapeutic mAbs which could assist in lowering the production costs and ultimately the cost of the final product (therapeutic drug) in the market.

References

Adalimumab: Highlights of Prescribing Information (FDA). (2018). https://www.accessdata.fda.gov/drugsatfda_docs/label/2018/125057s406lbl.pdf.

Adlersberg, J. B. (1976). The immunoglobulin hinge (interdomain) region. *La Ricerca in Clinica e in Laboratorio, 6*(3), 191–205.

Alessandri, L., Ouellette, D., Acquah, A., Rieser, M., Leblond, D., Saltarelli, M., ... Correia, I. (2012). Increased serum clearance of oligomannose species present on a human IgG1 molecule. *MAbs,* 4(4), 509–520.

Arata, Y. (1980). Proton nuclear magnetic resonance studies of human immunoglobulins: Conformation of the hinge region of the IgG1 immunoglobulin. *Biochemistry, 19*(22), 5130–5135.

Barbier, L., Declerck, P., Simoens, S., Neven, P., Vulto, A. G., & Huys, I. (2019). The arrival of biosimilar monoclonal antibodies in oncology: Clinical studies for trastuzumab biosimilars. *British Journal of Cancer, 121*(3), 199–210.

Behring, E. A., & Kitasato, S. (1890). Uber das Zustandekommen der Diphtherie-Immunitat und der Tetanus-Immunitat bei Thieren. Dtsch Med Wochenschr. *Deutsche Medizinische Wochenschrift (1946), 16*, 1113–1114.

Bennett, M. R. (2000). The concept of transmitter receptors: 100 years on. *Neuropharmacology, 39*(4), 523–546.

Bindon, C. I., Hale, G., Brüggemann, M., & Waldmann, H. (1988). Human monoclonal IgG isotypes differ in complement activating function at the level of C4 as well as C1q. *The Journal of Experimental Medicine, 168*(1), 127–142.

Bosch, F., & Rosich, L. (2008). The contributions of Paul Ehrlich to pharmacology: A tribute on the occasion of the centenary of his Nobel Prize. *Pharmacology, 82*(3), 171–179.

Boulianne, G. L. (1984). Production of functional chimaeric mouse/human antibody. *Nature, 312*(5995), 643–646.

Brezski, R. J., Vafa, O., Petrone, D., Tam, S. H., Powers, G., Ryan, M. H., et al. (2009). Tumor-associated and microbial proteases compromise host IgG effector functions by a single cleavage proximal to the hinge. *Proc Natl Acad Sci USA, 106*(42), 17864–17869.

Burnet, F. M. (1976). A modification of Jerne's theory of antibody production using the concept of clonal selection. *CA: A Cancer Journal for Clinicians, 26*(2), 119–121.

Cerny, T., Borisch, B., Introna, M., Johnson, P., & Rose, A. L. (2002). Mechanism of action of rituximab. *Anti-Cancer Drugs, 13*(Suppl. 2), S3–10.

Chothia, C., & Lesk, A. M. (1987). Canonical structures for the hypervariable regions of immunoglobulins. *Journal of Molecular Biology, 196*(4), 901–917.

Chothia, C., Lesk, A. M., Tramontano, A., Levitt, M., Smith-Gill, S. J., Air, G., et al. (1989). Conformations of immunoglobulin hypervariable regions. *Nature, 342*(6252), 877–883.

Cordoba, A. J., Shyong, B. J., Breen, D., & Harris, R. J. (2005). Non-enzymatic hinge region fragmentation of antibodies in solution. *Journal of Chromatography. B, Analytical Technologies in the Biomedical and Life Sciences, 818*(2), 115–121.

Dall'Acqua, W. F. (2006). Modulation of the effector functions of a human IgG1 through engineering of its hinge region. *Journal of Immunology, 177*(2), 1129–1138.

Davies, D. R., & Chacko, S. (1993). Antibody structure. *Accounts of Chemical Research, 26* (8), 421–427.

Edelman, G. M. (1959). Dissociation of y-globulin. *Journal of the American Chemical Society, 81*(12), 3155–3156.

Edelman, G. M., Cunningham, B. A., Gall, W. E., Gottlieb, P. D., Rutishauser, U., & Waxdal, M. J. (1969). The covalent structure of an entire γG immunoglobulin molecule. *Proceedings of the National Academy of Sciences of the United States of America, 63*(1), 78–85.

Ferrara, N. (2004). Discovery and development of bevacizumab, an anti-VEGF antibody for treating cancer. *Nature Reviews. Drug Discovery, 3*(5), 391–400.

Ferrara, N. (2016). Ten years of anti-vascular endothelial growth factor therapy. *Nature Reviews. Drug Discovery, 15*(6), 385–403.

Gearing, A. J. (2002). Selective cleavage of human IgG by the matrix metalloproteinases, matrilysin and stromelysin. *Immunology Letters, 81*(1), 41–48.

Generics and Biosimilars Initiative (GaBI) Online. (2018b). Biosimilars of rituximab. http://gabionline.net/Biosimilars/General/Biosimilars-of-rituximab.

Generics and Biosimilars Initiative (GaBI) Online. (2019). Biosimilars of trastuzumab. http://www.gabionline.net/Biosimilars/General/Biosimilars-of-trastuzumab.

Grundbacher, F. J. (1992). Behring's discovery of diphtheria and tetanus antitoxins. *Immunology Today, 13*(5), 188–190.

Honegger, A., & Plückthun, A. (2001). Yet another numbering scheme for immunoglobulin variable domains: an automatic modeling and analysis tool. *Journal of Molecular Biology, 309*(3), 657–670.

Hudis, C. A. (2007). Trastuzumab—Mechanism of action and use in clinical practice. *The New England Journal of Medicine, 357*(1), 39–51.

Idusogie, E. E., Wong, P. Y., Presta, L. G., Gazzano-Santoro, H., Totpal, K., Ultsch, M., et al. (2001). Engineered antibodies with increased activity to recruit complement. *Journal of Immunology, 166*(4), 2571–2575.

Jerne, N. K. (1955). The natural selection theory of Antibody formation. *Proceedings of the National Academy of Sciences of the United States of America, 41*(11), 849–857.

Jerne, N. K. (1974). Towards a network theory of the immune system. *Annals of Immunology, 125C*(1-2), 373–389, Paris.

Jerne, N. K., & Nordin, A. A. (1963). Plaque formation in agar by single antibody-producing cells. *Science (New York, N.Y.), 140*(3565), 405.

Jones, P. T., Dear, P. H., Foote, J., Neuberger, M. S., & Winter, G. (1986). Replacing the complementarity-determining regions in a human antibody with those from a mouse. *Nature, 321*(6069), 522–525.

Kaur, H. (2021). Characterization of glycosylation in monoclonal antibodies and its importance in therapeutic antibody development. *Critical Reviews in Biotechnology, 41(2),* 300–315 Available from https://doi.org/10.1080/07388551.2020.1869684.

Kaur, H., Bruno, J. G., Kumar, A., & Sharma, T. K. (2018). Aptamers in the therapeutics and diagnostics pipelines. *Theranostics, 8*(15), 4016–4032.

Kaur, H., & Yung, L. Y. (2012). Probing high affinity sequences of DNA aptamer against VEGF165. *PLoS One*, 7(2), e31196.

Kaur, H., Li, J. J., Bay, B. H., & Yung, L. Y. (2013). Investigating the antiproliferative activity of high affinity DNA aptamer on cancer cells. *PLoS One, 8*(1), e50964.

Kettleborough, C. A., Saldanha, J., Heath, V. J., Morrison, C. J., & Bendig, M. M. (1991). Humanization of a mouse monoclonal antibody by CDR-grafting: the importance of framework residues on loop conformation. *Protein Engineering, 4*(7), 773–783.

Keystone, E., & Haraoui, B. (2004). Adalimumab therapy in rheumatoid arthritis. *Rheumatic Diseases Clinics of North America, 30*(2), 349–364.

Kim, H. (1994). Dynamical structure of the hinge region of immunoglobulin G as studied by 13C nuclear magnetic resonance spectroscopy. *Journal of Molecular Biology, 236*(1), 300–309.

Klein, M., Haeffner-Cavaillon, N., Isenman, D. E., Rivat, C., Navia, M. A., Davies, D. R., et al. (1981). Expression of biological effector functions by immunoglobulin G molecules lacking the hinge region. *Proceedings of the National Academy of Sciences of the United States of America, 78*(1), 524–528.

Kohler, G., & Milstein, C. (1975). Continuous cultures of fused cells secreting antibody of predefined specificity. *Nature, 256*(5517), 495–497.

Kohler, G., & Milstein, C. (1976). Derivation of specific antibody-producing tissue culture and tumor lines by cell fusion. *European Journal of Immunology, 6*(7), 511–519.

Kunik, V., Good, R. J., & Chaudhury, M. K. (2013). The indistinguishability of epitopes from protein surface is explained by the distinct binding preferences of each of the six antigen-binding loops. *Protein Engineering, Design & Selection: PEDS, 26*(10), 599–609.

Landsteiner, K. (1945). The specificity of serological reactions, Cambridge, MA *Nature, 156*, 348–349.

Lefranc, M. P., Giudicelli, V., Ginestoux, C., Jabado-Michaloud, J., Folch, G., Bellahcene, F., et al. (1999). IMGT, the international ImMunoGeneTics information system. *Nucleic Acids Research, 33*(Database issue), D593–D597.

Lefranc, M. P., Giudicelli, V., Ginestoux, C., Jabado-Michaloud, J., Folch, G., Bellahcene, F., et al. (2003). IMGT unique numbering for immunoglobulin and T cell receptor variable domains and Ig superfamily V-like domains. *Developmental and Comparative Immunology*, *27*(1), 55–77.
Lis, H., & Sharon, N. (1993). Protein glycosylation. Structural and functional aspects. *European Journal of Biochemistry/FEBS*, *218*(1), 1–27.
Mallbris, L., Davies, J., Glasebrook, A., Tang, Y., Glaesner, W., & Nickoloff, B. J. (2016). Molecular insights into fully human and humanized monoclonal antibodies: What are the differences and should dermatologists care? *The Journal of Clinical and Aesthetic Dermatology*, *9*(7), 13–15.
Matsunaga, C. (1991). A 13C NMR study of the hinge region of a mouse monoclonal antibody. *Journal of Biomolecular NMR*, *1*(4), 379–390.
McCafferty, J., Griffiths, A. D., Winter, G., & Chiswell, D. J. (1990). Phage antibodies: Filamentous phage displaying antibody variable domains. *Nature*, *348*(6301), 552–554.
Molta, C. H. (2007). Etanercept. In W. H. Boehncke, & H. H. Radeke (Eds.), *Biologics in general medicine* (pp. 32–41). Berlin, Heidelberg: Springer.
Morrison, S. L. (1984). Chimeric human antibody molecules: Mouse antigen-binding domains with human constant region domains. *Proceedings of the National Academy of Sciences of the United States of America*, *81*(21), 6851–6855.
Musolino, A., Naldi, N., Bortesi, B., Pezzuolo, D., Capelletti, M., Missale, G., et al. (2008). Immunoglobulin G fragment C receptor polymorphisms and clinical efficacy of trastuzumab-based therapy in patients with HER-2/neu-positive metastatic breast cancer. *Journal of Clinical Oncology: Official Journal of the American Society of Clinical Oncology*, *26*(11), 1789–1796.
Natsume, A., Niwa, R., & Satoh, M. (2009). Improving effector functions of antibodies for cancer treatment: Enhancing ADCC and CDC. *Drug Design, Development and Therapy*, *3*, 7–16.
Norman, D. J. (1995). Mechanisms of action and overview of OKT3. *Therapeutic Drug Monitoring*, *17*(6), 615–620.
Padlan, E. A. (1990). On the nature of antibody combining sites: Unusual structural features that may confer on these sites an enhanced capacity for binding ligands. *Proteins*, 7(2), 112–124.
Pauling, L. (1940). A theory of the structure and process of formation of antibodies. *Journal of the American Chemical Society*, *62*(10), 2643–2657.
Pereira, N. A., Chan, K. F., Lin, P. C., & Song, Z. (2018). The "less-is-more" in therapeutic antibodies: Afucosylated anti-cancer antibodies with enhanced antibody-dependent cellular cytotoxicity. *MAbs*, *10*(5), 693–711.
Philippidis A. (2019). The top 15 best-selling drugs of 2018. https://www.genengnews.com/a-lists/top-15-best-selling-drugs-of-2018/.
Plosker, G. L. (2003). Rituximab: A review of its use in non-Hodgkin's lymphoma and chronic lymphocytic leukaemia. *Drugs*, *63*(8), 803–843.
Pohanka, M. (2009). Monoclonal and polyclonal antibodies production—Preparation of potent biorecognition element. *Journal of Applied Biomedicine*, 7(3), 115–121.
Porter, R. R. (1959). The hydrolysis of rabbit y-globulin and antibodies with crystalline papain. *The Biochemical Journal*, *73*, 119–126.
Potter, M. (1972). Immunoglobulin-producing tumors and myeloma proteins of mice. *Physiological Reviews*, *52*(3), 631–719.
Presta, L. G. (1997). Humanization of an anti-vascular endothelial growth factor monoclonal antibody for the therapy of solid tumors and other disorders. *Cancer Research*, *57* (20), 4593–4599.

Ricklin, D., & Lambris, J. D. (2007). Complement-targeted therapeutics. *Nature Biotechnology*, *25*(11), 1265–1275.

Roman, V. R. G., Murray, J. C., & Weiner, L. M. (2014). *Chapter 1—Antibody-dependent cellular cytotoxicity (ADCC). Antibody Fc* (pp. 1–27). Elsevier.

Roux, K. H. (1997). Flexibility of human IgG subclasses. *Journal of Immunology*, *159*(7), 3372–3382.

Salles, G., Barrett, M., Foà, R., Maurer, J., O'Brien, S., Valente, N., et al. (2017). Rituximab in B-cell hematologic malignancies: A review of 20 years of clinical experience. *Advances in Therapy*, *34*(10), 2232–2273.

Sarma, J. V., & Ward, P. A. (2011). The complement system. *Cell and Tissue Research*, *343* (1), 227–235.

Schirrmann, T., Meyer, T., Schütte, M., Frenzel, A., & Hust, M. (2011). Phage display for the generation of antibodies for proteome research, diagnostics and therapy. *Molecules (Basel, Switzerland)*, *16*(1), 412–426.

Schmidt, E. (2006). Rituximab in refractory autoimmune bullous diseases. *Clinical and Experimental Dermatology*, *31*(4), 503–508.

Schroff, R. W. (1985). Human anti-murine immunoglobulin responses in patients receiving monoclonal antibody therapy. *Cancer Research*, *45*(2), 879–885.

Schumaker, V. N. (1991). Dynamic aspects of antibody structure. *Molecular Immunology*, *28* (12), 1347–1360.

Sedrak, P., Hsu, K., & Mohan, C. (2003). Molecular signatures of anti-nuclear antibodies—Contribution of heavy chain framework residues. *Molecular Immunology*, *40*(8), 491–499.

Shawler, D. L. (1985). Human immune response to multiple injections of murine monoclonal IgG. *Journal of Immunology*, *135*(2), 1530–1535.

Shields, R. L., Namenuk, A. K., Hong, K., Meng, Y. G., Rae, J., Briggs, J., et al. (2001). High resolution mapping of the binding site on human IgG1 for Fc gamma RI, Fc gamma RII, Fc gamma RIII, and FcRn and design of IgG1 variants with improved binding to the Fc gamma R. *The Journal of Biological Chemistry*, *276*(9), 6591–6604.

Shih, T. (2006). Bevacizumab: An angiogenesis inhibitor for the treatment of solid malignancies. *Clinical Therapeutics*, *28*(11), 1779–1802.

Shinkawa, T., Nakamura, K., Yamane, N., Shoji-Hosaka, E., Kanda, Y., Sakurada, M., et al. (2003). The absence of fucose but not the presence of galactose or bisecting N-acetylglucosamine of human IgG1 complex-type oligosaccharides shows the critical role of enhancing antibody-dependent cellular cytotoxicity. *The Journal of Biological Chemistry*, *278*(5), 3466–3473.

Smith, G. P. (1985). Filamentous fusion phage: Novel expression vectors that display cloned antigens on the virion surface. *Science (New York, N.Y.)*, *228*(4705), 1315–1317.

Smith, M. A., Easton, M., Everett, P., Lewis, G., Payne, M., Riveros-Moreno, V., et al. (1996). Specific cleavage of immunoglobulin G by copper ions. *International Journal of Peptide and Protein Research*, *48*(1), 48–55.

Strober, B. E. (2005). The treatment of psoriasis with etanercept. *Seminars in Cutaneous Medicine and Surgery*, *24*(1), 28–36.

Tayar, J. H., Lopez-Olivo, M. A., & Suarez-Almazor, M. E. (2013). Adalimumab: 8 years of experience in rheumatoid arthritis. *International Journal of Clinical Rheumatology*, *8*(2), 165–184.

Thomann, M., Reckermann, K., Reusch, D., Prasser, J., & Tejada, M. L. (2016). Fc-galactosylation modulates antibody-dependent cellular cytotoxicity of therapeutic antibodies. *Molecular Immunology*, *73*, 69–75.

Valentine, R. C., & Green, N. M. (1967). Electron microscopy of an antibody-hapten complex. *Journal of Molecular Biology*, *27*(3), 615–617.
Van den Steen, P., et al. (1998). Concepts and principles of O-linked glycosylation. *Critical Reviews in Biochemistry and Molecular Biology*, *33*(3), 151–208.
van Oss, C. J., et al. (1986). Nature of the antigen-antibody interaction. Primary and secondary bonds: optimal conditions for association and dissociation. *Journal of Chromatography*, *376*, 111–119.
Vidarsson, G., Dekkers, G., & Rispens, T. (2014). IgG subclasses and allotypes: From structure to effector functions. *Frontiers in Immunology*, *5*, 520.
Winau, F., Westphal, O., & Winau, R. (2004). Paul Ehrlich—In search of the magic bullet. *Microbes and Infection/Institut Pasteur*, *6*(8), 786–789.
Winter, G., Griffiths, A. D., Hawkins, R. E., & Hoogenboom, H. R. (1994). Making antibodies by phage display technology. *Annual Review of Immunology*, *12*, 433–455.
Wu, T. T., & Kabat, E. A. (1970). An analysis of the sequences of the variable regions of Bence Jones proteins and myeloma light chains and their implications for antibody complementarity. *The Journal of Experimental Medicine*, *132*(2), 211–250.
Xiang, J., Sha, Y., Jia, Z., Prasad, L., & Delbaere, L. T. (1995). Framework residues 71 and 93 of the chimeric B72.3 antibody are major determinants of the conformation of heavy-chain hypervariable loops. *Journal of Molecular Biology*, *253*(3), 385–390.
Xiang, T. (2007). Structural effect of a recombinant monoclonal antibody on hinge region peptide bond hydrolysis. *Journal of Chromatography. B, Analytical Technologies in the Biomedical and Life Sciences*, *858*(1-2), 254–262.
Ying, T., Chen, W., Feng, Y., Wang, Y., Gong, R., & Dimitrov, D. S. (2013). Engineered soluble monomeric IgG1 CH3 domain: Generation, mechanisms of function, and implications for design of biological therapeutics. *The Journal of Biological Chemistry*, *288*(35), 25154–25164.
Yu, X., et al. (2014). *Chapter 10—Antibody glycosylation. Antibody Fc* (pp. 179–194). Elsevier Inc.

Further reading

General policies for monoclonal antibodies. (2009). https://www.who.int/medicines/services/inn/Generalpoliciesformonoclonalantibodies2009.pdf.
Generics and Biosimilars Initiative (GaBI) Online. (2018a). Biosimilars of trastuzumab. http://gabionline.net/Biosimilars/General/Biosimilars-of-trastuzumab.
Generics and Biosimilars Initiative (GaBI) Online. (2018c). Biosimilars of adalimumab. http://gabionline.net/Biosimilars/General/Biosimilars-of-adalimumab.
Generics and Biosimilars Initiative (GaBI) Online. (2018d). Biosimilars of Etanercept. http://www.gabionline.net/Biosimilars/General/Biosimilars-of-etanercept.
WHO. (2015). Model List of Essential Medicines (19th list).
https://www.nobelprize.org/prizes/chemistry/2018/press-release/.

CHAPTER 2

Physicochemical characterization of monoclonal antibodies

Harleen Kaur
Analytical Sciences, Aurobindo Biologics (Unit-XVII) (CurateQ Biologics), Hyderabad, India

2.1 Introduction

The physicochemical characterization of therapeutic monoclonal antibodies (mAbs) involves an extensive characterization of the drug molecule that is required to obtain regulatory approval and bring the new drug into the market. The physicochemical characterization is mandatory from the regulatory standpoint as it could influence the safety and efficacy of the drug. According to ICHQ6B guideline, the characterization of a biotechnological or biological products by appropriate techniques is necessary to establish the specifications which themselves are generated based on the data obtained from relevant development activities, lots used for preclinical/clinical studies (including consistency and clinical batches), stability studies, and analytical similarity (International Conference, 1999). The characterization includes determination of physical properties including purity and residual process impurities, primary structure, higher order structure (HOS), and biological activity of the desired product. The scope of this book chapter is limited to discussion about the analysis of physicochemical properties of monoclonal antibodies.

Like any therapeutic agents, monoclonal antibodies are also extensively characterized concerning their size, charge, structure, bioactivity, carbohydrate analysis and post translational modifications (PTMs). The PTMs including glycosylation have been discussed in detail in other chapters of the book.

The therapeutic molecules of increased complexity may require increased analytical characterization such as Etanercept where two N-glycosylation linked sites are present on TNF-α receptor and a third one is present in the Fc (fragment, crystallizable) region which is always located on the C-terminal domain of the mAb/fusion protein. In addition, Etanercept also contains 13 O-linked glycosylation sites which is unique to this molecule and not commonly found in monoclonal antibodies and thus, requires thorough investigation and

Monoclonal Antibodies
DOI: https://doi.org/10.1016/B978-0-12-822318-5.00007-7
Copyright © 2021 Harleen Kaur.
Published by Elsevier Inc. All rights reserved.

characterization of glycosylation (N- and O-linked) as glycosylation impacts the protein folding, solubility, pharmacokinetic behavior, clearance and potency of the drug and constitutes one-third of the Etanercept molecular weight (Beck et al., 2008; Beck et al., 2013; Montacir et al., 2018). The N- and O-linked glycosylation in mAbs and fusion proteins have been discussed in detail in Chapter 3, Glycosylation Analysis, of this book. The selection of analytical methods that are used in the physicochemical characterization of monoclonal antibodies highly dependents on the properties of the particular therapeutic moiety.

Over the years, it has become apparent that monoclonal antibodies develop various forms of heterogeneity in terms of size, charge and other parameters during manufacturing and storage. Monoclonal antibodies are subjected to enzymatic processes such as glycosylation or chemical degradation via mechanisms such as deamidation, oxidation, isomerization, glycation and fragmentation that result in the formation of charge or size variants or modified glycan forms (or glycan variants), thus modifying the antibodies and introducing heterogeneity in them. In the early stages of the drug development, every new batch of monoclonal antibody from upstream and downstream processing is subjected to extensive characterization particularly for size, charge and glycan heterogeneity to assess the batch to batch variability and individual batch quality. It is followed by process changes/process optimization depending on the obtained data that is required to ensure the batch consistency. The heterogeneity can also be introduced during the late stage manufacturing or storage of drug substance or drug product and thus it is indispensable to characterize the heterogeneity at this point as well and ensure lot to lot consistency for the batches particularly used in preclinical or clinical studies. As part of characterization, an in-house reference material also referred to as in-house reference standard (IRS) should be established from the manufacturing lots when no reference medicinal product (RMP) is available and should be used for physiochemical and biological testing/characterization of the pilot scale and manufacturing batches (International Conference, 1999). A Reference Standard, or Reference Material, is a substance prepared for use as the standard in an assay, identification, or purity test. Reference Standards play a critical role in calibrating and confirming the suitability of tests to assess product quality including characterization, comparability, lot release, and stability.

Specification of a drug substance/drug product are acceptance criteria that are proposed and justified by the manufacturer to be considered acceptable concerning quality for the intended use and is approved by regulatory authorities. Specification includes the list of analytical tests, references to the tests and acceptance criteria which are numerical limits, ranges or other criteria for each of the tests described (International Conference, 1999). Typical drug substance (DS) or drug product (DP) specifications include tests such as appearance, pH, osmolality, identity, purity and impurities (product- and process- related impurities), potency and quantity. The acceptance criteria is a set of criteria to which the DS or DP or in process samples should adhere to be confirmed acceptable for intended use. The acceptance criteria for each of the validated analytical methods employed for characterization of a product, in general are established by data generated from the development studies, stability studies, data demonstrating the consistency of the batches and preclinical/clinical studies. Including but not limited to establishing acceptance criteria, the data generated from aforementioned studies provides multitude of information about the drug manufacturing process and in-house product, allowing the process and product modifications to be performed accordingly. For the analytical methods according to pharmacopeias (compendial methods) such as determination of particle count, endotoxin, sterility/bioburden and volume in container assays should meet the acceptance criteria as set in the monograph/ pharmacopeia guideline.

2.2 Identification of critical quality attributes

The exhaustive physiochemical characterization of monoclonal antibodies and demonstration of comparability towards the reference product (in the case of biosimilars) is critical for the clinical evaluation and dossier submission. This could be achieved through detailed characterization of the product in terms of its primary structure, HOS, purity, glycosylation, potency, bioactivity and henceforth, developing deep understanding of the relationship between the specific physiochemical attributes and biological functionality (Putnam, Prabhu, Zheng, Subramanyam, & Wang, 2010; Li & d'Anjou, 2009). Protein characteristics are further defined in terms of the quality attributes of the molecule. Each of the quality attributes investigated by different analytical methods is defined as a critical quality attribute (CQA) or non-CQA based on the risk analysis. A proper identification of CQAs for the desired product is particularly important as it directly affects the safety and efficacy of the product and is critical to maintain within defined limits to ensure similarity, safety, potency and functionality. According to ICH Q8 R(2), a CQA is a physical, chemical, biological, or microbiological property or characteristic that should be within an appropriate limit, range, or distribution to ensure the desired product quality

(ICH Harmonised, 2009). CQAs are generally associated with DS, DP, excipients and intermediates. The CQAs of a product are derived from the quality target product profile (QTPP) and/or prior knowledge and can change with increasing knowledge of the product and process understanding. According to ICH Q8, QTPP is a prospective summary of the quality characteristics of a drug product that ideally will be achieved to ensure the desired quality, taking into account safety and efficacy of the drug product (ICH Harmonised, 2009). The QTPP should be considered as a development tool for which some target ranges may evolve during development, as further information on the reference product becomes available.

Production of mAbs through engineered mammalian or microbial cell expression system gives rise to heterogeneity and multiple variants through PTMs. Some of the PTMs like glycosylation, glycation, deamidation, oxidation, C-terminal lysine clipping and isomerization are common characteristics of monoclonal antibodies and may have a profound impact on their structure, stability and biological properties (Kaur, 2021b). From a regulatory standpoint, the location and identification of the kind of PTMs present in mAbs is crucial not only to understand the structure and function of the protein but also to identify the presence of undesired products, degradants or variant forms that could impact the overall purity, potency, and other quality attributes of the therapeutic product. According to ICH Q6B, an inherent degree of structural heterogeneity occurs in proteins due to the biosynthetic processes used by living organisms to produce them; therefore, the desired product can be a mixture of anticipated post-translationally modified forms (e.g., glycoforms) (International Conference, 1999). It is expected that the manufacturer should define the pattern of heterogeneity of the desired product and demonstrate consistency between lots used in preclinical and clinical studies. Not limited to antibody production, the heterogeneity could also be introduced during the drug substance or drug product storage and could have considerable influence on the product quality, potency and stability of the drug substance and drug product which makes it important to demonstrate lot to lot consistency of product heterogeneity and study the influence of environmental factors such as temperature, light, humidity and product quality with time (Jefferis, 2005).

2.2.1 Glycan heterogeneity

Glycosylation is one of the most common and well-studied PTM in therapeutic monoclonal antibodies which can result in remarkable heterogeneity in protein glycoforms with a potential to affect the safety/immunogenicity, efficacy/biological activity, and clearance [pharmacodynamics/pharmacokinetic property (PD/PK)] of the drug (ICH Harmonised, 2009; International Conference, 1999; Jefferis, 2005; Kaur, 2021a; Liu, 2015; Reusch & Tejada, 2015). Thus, glycan heterogeneity is a critical parameter to monitor during different stages of upstream and downstream operations and should be characterized for the carbohydrate content, structure of the carbohydrate chain, oligosaccharides and glycosylation sites present on the antibody. The method development activity of glycan analysis if done with a chromatographic method must evaluate the proper resolution of different glycan peaks, optimization of the mobile phase gradient, run time, column temperature and precision of the developed method. In lot release assay (LRA) testing and analytical similarity assessment, glycan heterogeneity is characterized by means of relative quantitation of different glycan forms such as amounts of afucosylation (G0 + G1 + G2), galactosylation (G1 + G2 + G1F + G2F), high mannose (M4 + M5 + M6 + M7 + M8), and non-glycosylated heavy chain (NGHC) and comparing the levels of respective glycan forms in the test samples to the reference standard (RMP/IRS) (Jung et al., 2014; Lee et al., 2018). In general, the physiochemical characterization of the glycan variants in the monoclonal antibody against the reference product (RMP/IRS) mainly shows G0F and G1F as the major glycoform species, while the less abundant species include G2, M5, and M6 which vary from antibody to antibody. The obtained relative abundance glycan species must be comparable and demonstrate similar heterogeneity pattern to the reference sample standard (RMP/IRS).

The afucosylated glycoform is linked to enhance the binding of the Fc domain of the antibody to FcγRIIIa and improves the antibody dependent cell mediated cytotoxicity (ADCC), whereas high degree of galactosylation can trigger the activation of the complement pathway by increasing the complement component 1q (C1q) binding and complement dependent cytotoxicity (CDC) activity. This was observed in Rituximab and Campath-1H antibodies (Boyd, Lines, & Patel, 1995; Hodoniczky, Zheng, & James, 2005; Shields et al., 2002). Hodoniczky et al. showed that degalactosylation of remodeled Rituximab reduces CDC by approximately half compared to the unmodified antibody and the addition of bisecting N-acetylglucosamine (GlcNAc) demonstrated 10-fold increase in the ADCC with minimal impact on the CDC in both Rituximab and Herceptin (Hodoniczky et al., 2005). These findings are in agreement with the reported literature in which the removal of galactose in another monoclonal antibody Campath-1H resulted in reduction in CDC activity by 35%–50% with no effect on the ADCC (Boyd et al., 1995). High mannose (Mannose 5 glycan- M5; Mannose 6 glycan-M6 and so on) bearing antibodies are cleared more rapidly in humans than antibody with other glycans indicating the high mannose content of therapeutic MAbs as an important quality attribute as it can impact the pharmacokinetic behavior and efficacy of the antibody (Goetze et al., 2011). Goetze et al. used one IgG1 and three IgG2 therapeutic antibodies to demonstrate that Man 5 glycans levels decreased rapidly as function of circulation time compared to other glycan form and thus concluded that Fc containing high mannose glycans are cleared faster in humans than other glycan forms (Goetze et al., 2011).

Sialylation is a protein modification which involves addition of charged glycans referred to as sialic acid (N-acetyl neuraminic acid, NANA or N-glycol neuraminic acid, NGNA) to the non-reducing end of the oligosaccharide chain. Fc glycan sialylation is known to reduce the ADCC activity by inhibiting the binding of the Fc domain to FcγRIIIa receptor on the natural killer (NK) cells and since ADCC is one of the therapeutic mechanism for mAbs, the sialylation content should be characterized and its influence on the binding and ADCC should be determined as part of quality attribute assessment (Scallon, Tam, McCarthy, Cai, & Raju, 2007). The sialic acid analysis involves quantifying the NANA levels in monoclonal antibody sample and is expressed as mmole/mole of protein. The aglycosylation level in the antibody is determined by measuring the content of the NGHC species in the antibody. It is known to affect the antibody's bioactivity and effector function (Reusch & Tejada, 2015). The NGHC levels are analyzed by capillary gel electrophoresis (CE-SDS) in reduced conditions and comparing the test sample results to the reference standard. The CE-SDS method has been discussed in Section 2.2.2.3 of this book chapter.

2.2.2 Size heterogeneity

Recombinant monoclonal antibody commonly generates size variants such as high molecular weight (HMW) species/aggregates and low molecular weight (LMW) species/fragments/degradants during manufacturing and storage. HMW and LMW are also referred to as product related impurities (International Conference, 1999). The presence of LMW and especially HMW have the potential to elicit immune response and affect immunogenicity and potency of the finished product (Ratanji, Derrick, Dearman, & Kimber, 2014; Rosenberg, 2006; Vlasak & Ionescu, 2011). As a result, the HMW and LMW are considered to be CQAs and their levels are routinely monitored as part of drug development process, lot release testing, stability testing and characterization activities for both the DS and DP. The analytical methods that are commonly used for detection and quantification of size heterogeneity in monoclonal antibodies include (but not limited to) size exclusion chromatography (SEC), sodium dodecyl sulfate-polyacrylamide gel electrophoresis (SDS-PAGE), and capillary electrophoresis-sodium dodecyl sulfate (CE-SDS). These analytical methods will be discussed in the following sections.

2.2.2.1 Size exclusion chromatography

SEC is a critical method to monitor the amount of aggregates/high molecular weight (HMW) species in mAbs and is used to assess product purity by distinguishing monomeric IgG from aggregates and fragments (LMW) in non-reducing conditions. This method is based on the principle of differential sieving where the HMW pass through the voids between the pores in the stationary phase while the LMW enters the pores experiencing high retention time compared to HMW and elute through the column based on their molecular weight/size. In order to improve the resolution and method robustness in SEC, several parameters have to be evaluated as part of SEC method development. They include mobile phase composition, presence or absence of sodium chloride (NaCl) in the mobile phase (also referred to as ionic strength) and the type of column (in terms of column length, pore size and column separation range) depending on the size of the molecule, flow rate and run time.

A typical SEC profile shows clear separation between the monomer, HMW, and LMW species. The purity of the product analyzed via SEC is based on the relative product monomer content with respect to HMW and LMW. The obtained levels of size variants must be comparable to the primary reference standard (RMP/IRS). An analytical similarity assessment of Rituximab biosimilar CT-P10 drug product to the RMP demonstrated slightly higher levels of the monomer content (0.3%–0.5%) in the in-house product while the level of HMW were relatively lower compared to the US- and EU-Rituximab in SEC analysis (Lee et al., 2018). Negligible levels of LMW were observed (less than 0.5%). Another comparability study between a biosimilar Trastzumab – Probiomed and its reference product showed the levels of monomer content analyzed by SEC-UPLC (ultra) performance liquid chromatography were comparable for three batches of the in-house product versus the RMP and the contents of the aggregates and fragments were below the limits established by the United States Pharmacoepia (López-Morales et al., 2015).

Size variants are part of the stability indicating quality attributes and increase in the course of the shelf life of the product. Stability indicating attributes are assessed by comparing the degradation profiles of the in-house product and RMP. As part of the analytical similarity exercise, Seo et al. thermally degraded the US- and EU- Bevacizumab and in-house Bevacizumab biosimilar product ABP 215 by incubating at 50°C for 14 days. The SEC analysis demonstrated similar degradation profiles between Bevacizumab and ABP 215 for HMW, monomer and LMW species (Seo et al., 2018). SEC alone or in conjunction with electrospray ionization (ESI)/mass spectrometry (MS) has shown promising potential to identify and quantify size variants in the recombinant monoclonal antibodies. Haberger and colleagues employed SEC-UPLC coupled with native mass spectrometry (MS) to investigate the presence of charge variants in an in-house bispecific antibody during early stages of drug development process (Haberger et al., 2016). The native SEC-UV/MS method was able to detect and resolve the aggregates (dimers and trimers), monomer and the fragments in stability samples stored at 5°C for 24 months and accelerated stability samples incubated at 40°C for 3 months. The real time stability samples stored at 5°C showed formation of dimers whereas truncated species were detected in accelerated temperature induced samples. Furthermore, the native SEC-UV/MS method was able to resolve and detect the formation of aggregates and unbound light chain (LC) fragments in early stages of purification process. They eventually got removed in the subsequent purification steps. The findings from this study indicates the potential of the SEC-UV/MS method identification and in depth analysis of the size variants in the antibody during the early development phase.

2.2.2.2 Sodium dodecyl sulfate-polyacrylamide gel electrophoresis

SDS-PAGE is a qualitative method used to monitor overall size fragmentation in the monoclonal antibody. The principle of SDS-PAGE involves the electrophoretic movement of the molecules and separates the samples based on their molecular weights. The lipophilic tail of the SDS binds non-covalently to the protein, thus conferring negative charge and masking the intrinsic charge of the protein. The rate of migration on the gel is inversely proportional to the molecular weight of the protein sample. Following some examples of the employment of SDS-PAGE are given.

Peg-Filgrastim is a covalent conjugate of filgrastim (a recombinant human granulocyte colony stimulating factor) and monomethoxypolyethylene glycol (mPEG). Filgrastim is produced in *E. coli* as a single, nonglycosylated, polypeptide chain containing 175 amino acids and with a molecular mass of 18,800 Da. Pegylated granulocyte colony stimulating factor (GCSF) is manufactured by conjugating a 20,000 Da monomethoxypolyethylene glycol propionaldehyde (mPEG-ALD) to the N-terminal of methionine of GCSF having a total molecular mass of 38,800 Da. In order to characterize the product related impurities in pegfilgrastim, Shekawat and colleagues employed nonreducing SDS-PAGE (4%–20% pre-cast gel) coupled with silver staining and assessed the concentrated and purified cation exchange chromatography (CEX) impurity pre-peaks (Pre-peak 1, 2 and 3 referred in the paper) and the main peak collected from CEX HPLC analysis (Purity >95%) (Shekhawat et al., 2019). The internal reference standard of pegfilgrastim and filgrastim was loaded at 0.015 μg (0.15%), 0.050 μg (0.5%), 0.15 μg (1.5%) and 10.00 μg (100%) concentrations. The absence of band near 15 kDa in SDS-PAGE indicate the absence of free filgrastim in the impurity samples. Presence of major band between 75–100 kDa in all the pre-peak sample lanes indicate the possibility of the dipegylated species and another major band in pre-peak 3 impurity sample at size corresponding to the main band for pegfilgrastim (~50 kDa) indicate the possibility of the monopegylated positional isomers in the impurity samples (Peak 3). The pegfilgrastim main peak was observed at 50 kDa.

In another study, Hawe et al. have used SDS-PAGE to understand the nature of aggregates formed in thermal stressed and freeze-thawed samples of a monoclonal antibody (Hawe, Kasper, Friess, & Jiskoot, 2009). Under nonreducing conditions, in addition to the one major band formed at 150 kDa, a band of large aggregates was formed at the top of the gel in thermal stressed samples. However, this band of aggregates disappeared in the reducing conditions indicating they were linked by disulfide bonds and were covalent in nature. No additional bands were observed in freeze-thawed samples indicating the aggregates formed in these samples were noncovalent. There is a clear evidence from the aforementioned studies that SDS-PAGE is a valuable technique for qualitative analysis of aggregates which are potentially immunogenic and impact the potency of the product. However, the disadvantages of this technique is that it is labor intensive, time consuming, nonquantitative and shows poor band resolution for samples with less than 10 kDa molecular weight difference.

2.2.2.3 Capillary electrophoresis-sodium dodecyl sulfate

The most common methods for size variants characterization are SEC and SDS-PAGE. SEC methods suffer from limitations such as low sensitivity and resolving power for low molecular weight species (LMW) whereas SDS-PAGE although fairly sensitive with silver staining has some major drawbacks which includes inaccurate quantification, the use of toxic reagents, manual labor and long run time. CE-SDS, showing greater reproducibility, enhanced resolution, high sensitivity and accurate quantification of LMW, is able to overcome the apparent drawbacks observed in SEC and SDS-PAGE. CE-SDS employs narrow bore, bare fused silica capillary (effective length 20.2 cm and internal diameter 50 μm) to separate the denatured protein molecules under reduced and nonreduced conditions based on their molecular weight. The native charge of the protein molecules is masked by the negative charge of the SDS resulting in separation according to differences in protein molecular weight. Small molecules will move faster than large molecules. When the voltage is applied (reversed polarity), the negatively charged molecules (SDS-protein complexes) move towards the anode and the signal is detected by the photodiode array or laser induced fluorescence (LIF) detection system. Often, CE-SDS is used to quantitate the LMW (1H, 1L, 1H1L, 2H and 2H1L) in nonreduced conditions and under reduced conditions, the test sample and the system suitability sample are treated with β-mercaptoethanol prior to injection and are used for quantitation of the NGHC. CE-SDS technique has demonstrated to detect NGHC as low as less than 2% (Ma, 2005). In order to enhance the reliability and reproducibility of CE-SDS assay, sample preparation is one of critical parameters that needs to be optimized because the sample is exposed to high temperature which can lead to fragmentation. The CE-SDS method development exercise have to include optimization of the sample buffer and alkylating agent concentration sample incubation temperature and incubation time. It is known that alkylation of proteins with iodoacetic acid (IAA) or iodoacetamide (IAM) is more efficient at higher pH ($pH > 8.0$) due to formation of cysteine in anionic form (Lundell & Schreitmüller, 1999). Salas et al. investigated the effect of the alkylation with IAA and IAM on the fragmentation of the monoclonal antibody (Salas-Solano et al., 2006). Samples were prepared by mixing the mAb samples with Tris-HCl sample buffer (pH 9.0) with SDS solution (final SDS concentration 1% wt./wt.) with varying amounts of IAM or IAA (10, 40, 80, and 100 mM). Compared to the control sample (without IAA/IAM), a significant decrease in the corrected peak area of the LMW was observed in the samples treated with 40 mM IAM in CE-SDS analysis. No further decrease in the fragmentation was observed at concentrations higher than 40 mM. Similar results were obtained with IAA. However IAM was the preferred choice as IAA was less stable as indicated by appearance of a yellow coloration in a few minutes of sample preparation. The aggregation level was the same in the alkylated and non-alkylated samples (0.2%). In the case of reduced samples which were treated with 1 M dithiothreitol (DTT), no significant effect of alkylation was observed compared to the control samples indicating the importance of alkylation in preventing fragmentation only in nonreduced conditions. In the same study, the authors also examined the effect of incubation temperature and time on mAb fragmentation in the presence and absence of the alkylating agent. In the absence of IAM, the corrected peak area (CPA) percentage of the intact antibody decreased from 97% to 90% with the increase in temperature (45°C, 60°C, and 90°C) and time, however, in the presence of IAM in the same temperature conditions, the thermal induced fragmentation was suppressed with CPA percentage of intact antibody remained at 97% even after 20 min of incubation at those temperatures. This data demonstrates that alkylation of monoclonal antibody samples at the heating step of protein-SDS complex at different temperatures improves the robustness and accuracy of the CE-SDS method. The effect of the sample buffer was determined by reconstituting the mAb sample in different buffers in the presence and absence of the 40 mM IAM. The CPA percentage of the intact antibody peak was highest at pH 6.5 and decreased with the increase in the pH for samples incubated with IAM. No change in the CPA percentage was observed at decreasing pH levels below 6.5, however, peak broadening was observed for the intact peaks and its size variants which could be attributed to the lower stability of SDS-protein complexes in acidic conditions. Overall, the results from Salas et al. demonstrated that thermal induced fragmentation in mAb can be reduced by systematic optimization of the sample preparation conditions prior to CE-SDS analysis. Considering fragmentation and nonglycosylation can have negative impacts on the potency and effector function of the antibody, Lee et al. employed CE-SDS to quantitate the levels of intact peak and NGHC in Rituximab and its biosimilar product CT-P10 (Lee et al., 2018; Reusch & Tejada, 2015; Vlasak & Ionescu, 2011). The results from analytical similarity assessment demonstrated that compared to Rituximab, 2% higher levels of intact antibody were observed in biosimilar product CT-P10 in nonreduced CE-SDS using 250 mM IAM. On the other hand, in reduced conditions, the CT-P10 showed slightly higher levels of NGHC (mean value 0.80%) compared to that EU-Rituximab (mean value 0.43%) and US-Rituximab (mean value 0.47%). Taking the results into account, the in-house product CT-P10 has marginally better quality profile in terms of purity/impurity compared to the RMP.

2.2.3 Charge heterogeneity

In addition to glycan distribution and size heterogeneity, charge heterogeneity is an important attribute concerning antibody characterization. Antibody charge variants have gained considerable attention and are a CQA as they can influence the stability and product quality of antibody therapeutics. Common PTMs which result in alteration of charge characteristic include *deamidation* of asparagine and glutamine residues. It involves loss of amide group resulting in formation of asparatic acid and glutamic acid, respectively and addition of negative charge unit on the mAbs and cause decrease in the pI value thus resulting in generation of acidic variants.

C-terminal lysine clipping leads to removal of an amine group bearing C-terminal amino acid which results in loss of one positive unit and formation of acidic variants. *Cyclization of glutamine/glutamate to pyroglutamate (pGlu)* involves loss of positively charged primary amine generating basic variants and *isomerization of aspartate to isoaspartate* makes the antibody slightly more acidic (Jefferis, 2016) [Table 2.1]. The case studies demonstrating the effect of the PTMs on antibody's binding affinity and biological activity will be discussed in detail later in this section.

Understanding the chemical nature of the acidic, basic and main species and their effect on the antibody's physicochemical properties is critical for the process design and formulation development. The analytical methods that are commonly used as part of early stage drug development, lot release testing, stability testing, characterization and analytical similarity exercise in laboratories for detection of charge variants include CEX, anion exchange chromatography (AEX), capillary isoelectric focusing (cIEF) and capillary zone electrophoresis (CZE). In chromatography based

Table 2.1 List of post translational modifications generating acidic and basic variants in monoclonal antibodies.

	Type of post translational modification	Modification location on the antibody surface	Change in molecular weight (Da)
Formation of Acidic Species	Deamidation	Variable and constant regions. Frequently in the CDR regions	1 Da shift (positive shift)
	Glycation	Lysine residue of the heavy chain or LC or both	162 Da shift (hexose sugar, monoglycated, positive shift)
	Sialylation	Asn residue of the CH2 domain of the heavy chain	291 Da shift (one sialic acid, positive shift)
	Oxidation	Pre-dominantly methionine residue in the Fc region. Also, observed in trypthophan, tyrosine, cysteine and histidine	16 Da shift (positive shift)
Formation of Basic species	C-terminal lysine	C-terminal of the heavy chain	128 Da shift (positive shift)
	Cyclization of glutamic acid/N-terminal glutamine to pyroglutamate (pE)	Heavy chain or LC or both	17 Da shift (negative shift)
	Isomerization of aspartate to isoaspartate	Frequently in the CDR regions	No mass shift
	Succinimide (common intermediate for Asp deamidation and Asp isomerization, mac_vp="28"/ >Loss of -COOH)	Heavy chain or LC or both	17 Da shift

LC, Light chain.

methods, the acidic and basic fractions are defined based on their retention time, while in cIEF, the charged species are separated based on the *overall charge difference or apparent isoelectric point (pI)*. CEX and AEX separate the molecule based on their net negative charge, where CEX uses a negatively charged resin with an affinity for molecules with net positive charge and AEX uses a positively charged resin with an affinity for molecules with net negative charge. As a consequence, in CEX, the acidic peaks elute before the main peak and the basic peaks elute after the main peak. In contrast with AEX, the basic peak elute earlier than the main peak which is followed by elution of the acidic peak. Amongst the two different ion exchange chromatography (IEX) modes, CEX is the best approach for both the qualitative and quantitative evaluation of charge variants in mAbs and uses either salt based gradient or pH based gradient for the elution (Fekete, Beck, Fekete, & Guillarme 2015a, 2015b; Salas-Solano et al., 2011). For most of the mAbs, the salt gradient approach is preferred as it results in better variant resolution compared to the pH gradient approach. In salt gradient approach, the molecule is eluted by increasing the salt concentration to increase the competition between the buffer ions and proteins for positively charged residues on the ion exchange resin. The pH of the mobile phase buffers in the salt gradient mode in CEX is between 6 and 7.5. because the pI of the monoclonal antibodies is between 7 and 9 (basic) and lower pH of the buffer increases the positive charges on the antibody resulting in longer retention time in the column. On the other hand, the pH gradient approach uses binding and elution buffer systems at different pH (range is between 4—10) while keeping the salt concentration constant. The development strategy for a CEX method must consider the optimization of the mobile buffer pH and salt concentration gradient. They strongly affect the ionization state of the antibody and eventually the final chromatographic profile. Additionally, column temperature (as lower temperature in CEX favors better IEX separation), column load, sample dilution, caroboxypeptidase-B (CPB) digestion (protein:enzyme ratio and incubation time) and detection wavelength should also be optimized.

Lee and colleagues employed weak cation exchange chromatography using gradient NaCl elution (200 mM NaCl, pH 6.0) at 214 nm for investigating the analytical similarity between innovator Rituximab and in-house biosimilar product (CT-P10) (Lee et al., 2018). Compared to RMP (EU- and US-Rituximab), highly basic variants were identified in CT-P10 due to high levels of N-terminal glutamine variants in the light chain (peak fraction analyzed by MS). However, N-terminal glutamine variants have no impact on potency, safety, or biological activity.

In contrast to CEX and AEX where the molecules are separated based on their surface charge, the capillary methods such as cIEF and CZE separate based on the net charge of the complete molecule. cIEF serves as an orthogonal technique to IEC for charge variant analysis that works on the principle of separating the protein molecules based on their pI in a capillary. A mixture of ampholytes (pH 3−10) migrate to their pH zones which correspond to their pI values and is used to establish a pH gradient in a capillary. This self-regulating process is referred to as focusing. During the focusing step, the neutral-coated capillary ends are immersed in anolyte and catholyte solution (H_3O^+/OH^-) and the applied voltage causes the pI markers, pH gradients (or ampholytes) and proteins to migrate towards their pH range. The migration ends when the pH of the molecule is equal to its pI. In other words, high voltage focusing establishes a pH gradient in the capillary, which allows proteins/peptides to focus at its corresponding pI. Compared to CEX, collection of fractions is difficult in cIEF and is generally used for monitoring of variants and not for their characterization. cIEF employs charged couple device (CCD) which is a highly sensitive photon detector for imaging purpose. The CCD is divided into a large number of light-sensitive small areas (known as pixels) which is used to build up an image of the scene of interest. A photon of light which falls within the area defined by one of the pixels converts into one (or more) electrons and the number of electrons collected is directly proportional to the intensity of the scene at each pixel. When the CCD is clocked out, the number of electrons in each pixel are measured and the scene can be reconstructed. For real time monitoring, the UV absorbance is measured at 280 nm along with native fluorescence generated from the aromatic amino acids (excitation wavelength: 280 nm and emission wavelength: 450 nm).

In 2011, an interlaboratory comparison study led by Salas-Solano et al. was carried out to study the charge heterogeneity of mAbs by evaluating the precision and robustness of cIEF (Salas-Solano et al., 2011). This work was a collaboration of 11 laboratories located in North America and one in Europe to study the robustness of cIEF in terms of pI and determination of charge variants in mAbs. The PA800 CE instrument equipped with UV detector from Beckman Coulter was employed by all the participants for the study. The study was performed using the same test mAb sample by all laboratories. The experiments were conducted in two scenarios. In one scenario, the test sample was prepared twice using the same lot of all the cIEF reagents including the ampholytes (Lot A1) and each sample was injected six times. In the second scenario, two different sample preparations were analyzed using two different lots of ampholytes (Lot A1 and Lot A2) but the same lot of the rest of the reagents. The obtained data shows a very good precision for pI between the laboratories with relative standard deviation (RSD) percentage of $\leq 0.5\%$ for ampholyte A1 and $\leq 0.4\%$ for ampholyte A2. The quantitative analysis of the charge variants across different laboratories, different analysts, different lots of ampholytes and multiple instruments showed RSD percentage for total peak area percentage was $\leq 9\%$ using ampholyte lot A1 and $\leq 8.4\%$ using ampholyte lot A2. The removal of outlying data improved the precision with RSD percentage for total peak area percentage $\leq 4.4\%$ using ampholyte lot A1 and $\leq 3.6\%$ using ampholyte lot A2, respectively. Overall, this exercise demonstrated the robustness and reproducibility of the cIEF method across different laboratories, different analyst, and different equipment for the same model number and different ampholyte lots.

In order to improve the cIEF method, following factors that could affect the focusing and separation of the proteins inside the capillary in the electric field have to be optimized:

1. Sample concentration: When the protein concentration is too high, proteins with low solubility will precipitate resulting in poor reproducibility (He, Que, & Mo, 2009). Therefore an antibody sample with high concentration must be diluted prior to the measurement. However, the protein concentration should be high enough so that sensitivity is adequate for detection and quantification of minor charge variants.
2. Focusing time: The dimensions of the capillary length and diameter have direct impact on the focusing time and voltage. Greater capillary length and/or diameter require longer focusing time to separate. This might affect the resolution and results in longer analysis time. Optimization of the focusing time is important because if focusing time is too low, sample molecules would not be separated well. If the focusing time is too long, protein precipitation or aggregation might occur (He et al., 2009). He et al. investigated the effect of different focusing times on charge separation in mAbs (referred to as mAb2 and mAb3 in the paper) at a constant voltage of 3000 V and one minute of pre-focusing at 1500 V was applied to prevent possible protein degradation caused by salt generated high current and heat (He et al., 2009).

The results demonstrated optimal focusing time range of 4−8 min for mAb2 whereas for mAb3 the focusing time range was broad with complete focusing achieved from 6 up to 18 min.

3. Anodic and cathodic stabilizers: The anodic and cathodic stabilizers are added to keep the pH gradient stable and prevent the loss of sample and ampholytes (H_3PO_4 and NaOH) at the anode and cathode. They are also referred to as "sacrificing agents" and are first to be consumed when the pH gradient decays. L-arginine (pI 10.8) is commonly used as the cathodic stabilizer and iminodiacetic acid (IDA, pI 2.2) is used as the anodic stabilizer. High concentration of the stabilizers allows the pI markers to resolve properly.
4. Carrier ampholytes: Carrier ampholytes are a mixture of polycarboxylic amino acids with pI ranging from 2.5−10. The electrophoretic characteristics of the vendor varies from vendor to vendor and batch to batch. Adequate concentrations of ampholytes is necessary to achieve a stable pH gradient. However, a high concentration of ampholyte can generate a noisy background or less stable baseline which could be due to Joule heating (He et al., 2009).
5. Urea: Urea is added as a solubility enhancer to increase the protein solubility which is typically low at or near its pI. Low solubility causes proteins to precipitate and aggregate resulting in lower cIEF separation reproducibility. Urea at concentration ≤ 4 M denature the non-covalent bonds that leads to protein aggregation and precipitation. Mack et al. demonstrated the significant improvement in peak resolution with usage of urea (0−3 M) in cIEF analysis of a mAb (Mack, Cruzado-Park, Chapman, Ratnayake, & Vigh, 2009). However, Suba et al. developed a cIEF method where no use of salt or urea resulted in better peak resolution compared to use of 25 mM Tris solution or 4 M urea added to the mAb sample solution, thus minimizing the complexity of the sample preparation (Suba, Urbányi, & Salgó, 2015).
6. Methylcellulose (MC) concentration: MC serves as a sieving matrix to reduce the electroosmotic flow (EOF) and a dynamic coating agent to prevent the adsorption of the proteins to the inner capillary surface. He et al. investigated the effect of different MC concentrations (0.25%. 0.35%, and 0.45%) on total area and main peak height of mAb in cIEF analysis (Garza et al., 2013). The results showed decrease in total area and main peak height with increase in MC concentration. High MC concentration is required to achieve better sensitivity. However, since MC is a polymer with high viscosity, high MC

concentration increases instrument related problems such as clogging of transfer line, and difficulty in pipetting which can led to reducing the precision of the instruments. Considering the associated challenges and obtained results, 0.35% MC was found to be the optimized concentration that could produce adequate sensitivity and precision.

Capillary zone electrophoresis (CZE) is an alternative technique used for analysis of charge heterogeneity of mAbs and provides complementary information. The molecules migrate with different speeds under the influence of an electric field resulting in earlier elution of basic variants, followed by the main peak and late elution of the acidic variants. The separation is based on different electrophoretic mobilities (charge to mass ratio), which result from different ratios of charge and hydrodynamic radius of the analytes when they are present in an electric field. In addition, the separation is dependent on the EOF which is a unidirectional flow created by formation of ionic bilayer from the separation buffer The separation is mostly performed using a permanently coated capillary to suppress the EOF and minimize the adsorption of the protein molecules to the inner capillary surface. If permanent coated capillaries are not available, a dynamic coating needs to be created in order to prevent the adsorption of the protein to the capillary surface. In order to prevent the protein adsorption, three additives that commonly constitute the CZE running buffer (pH 5.7) include (Garza et al., 2013)

1. Epsilon amino caproic acid (EACA): EACA binds with silanol groups on the capillary surface and form ion pairs with protein molecules which reduce their net charges and decreases the interactions with silanol groups (Garza et al., 2013).
2. Triethylenetetramine (TETA): This organic modifier forms a dynamic coating on the capillary wall and competes with proteins for interactions with silanol groups. In addition, it helps to improve the peak shape and resolution at higher pI values (Garza et al., 2013).
3. Hydroxypropyl methyl cellulose (HPMC): HPMC serves as a viscosity enhancer and adsorbs to the neutral surfaces to suppress interactions.

The CZE application uses UV detection at 214 nm or LIF detection for a better sensitivity and operates at a separation voltage of 30 kV. He and colleagues utilized an uncoated bare, fused silica capillary column in CZE method for the detection and separation of charge variants in mAbs. The results of the work showed that EACA present in the running buffer in combination with 0.1 N HCl flushing (referred to as acid conditioning) between the runs had significant impact on the peak resolution and suppression of EOF inside the capillary (He et al., 2010). Better separation efficiency was observed with 600 mM EACA concentration and introduction of 0.1 N HCl flushing between the injections reduced the peak broadening particularly in the basic variants. It was hypothesized that conditioning the capillary with 0.1 N HCl reduces the adsorption of the protein to the inner surface of the uncoated capillary surface thus allowing the adsorption of nonionic HPMC to the neutral surface. In addition, concentrated EACA can interact with residual silanol groups that are not covered by HPMC. The developed method demonstrated reproducibility with RSD percentage of less than 10% for different charge variants and migration time when the analysis was performed by two analysts using the same sample on three consecutive days. The work also showed that a longer, uncoated capillary (40 cm) can result in better separation of variants compared to the coated capillary which is less economical. Another study from the same group demonstrated that optimization of the pH of the running buffer, concentrations of EACA and TETA can have significant effect on the separation of the peaks and improve the peak shape. In addition, the work also showed the potential of CZE as a stability indicating method where an IgG mAb was stored at different temperatures (−20°C to 40°C). The main species decreased significantly from 25°C to 40°C, whereas the acidic and basic species increased with increase in temperature due to deamidation and formation of cyclic N-pyroglutamate, respectively (He, Isele, Hou, & Ruesch, 2011).

Hintersteiner et al. employed weak cation exchange chromatography (pH gradient elution method) to separate acidic, main (neutral variants), and basic charge variants of a chimeric antibody on a preparative scale and gather insights about the effect of the charge heterogeneity on the antibody binding and its effector functions (Hintersteiner et al., 2016). The data from the work revealed that basic antibody charge variants demonstrated higher binding affinity for the FcγRIIIa and FcγRn receptors and significantly better performance for ADCC and CDC assay compared to the acidic variants. Additionally, the authors also investigated the presence of other post translational modifications which could impact the structural integrity and stability of the antibody. The glycan data showed higher amounts of sialylated glycan forms in earlier fractions (acidic charge variants) while later eluting fractions contained afucosylated and high mannose glycan type structures. The peptide mapping data from the first basic fraction (referred to as B1 in the paper) indicated the presence of N-terminal pyroglutamate formation and C-terminal lysine clipping and, the later eluting basic fractions (referred to as B2 in the paper) indicated the presence of oxidation of a methionine residue in the complementarity determining region (CDR) region, However no significant impact on the binding affinity was observed from either of the detected modifications and it was speculated to result in conformational changes in the antibody.

In another charge variant based characterization study, the researchers examined the effects of isolated acidic, main, and basic peak fractions of an Avastin/Bevacizumab biosimilar product on the in vitro properties and pharmacokinetic behavior of the drug and compared the obtained results to the innovator molecule (Zhao et al., 2016). Strong cation exchange (SCX) chromatography was employed for isolation of the charge variants followed by buffer exchange and purity analysis by weak cation exchange (WCX) chromatography. A salt based gradient was used for the elution with mobile phase containing 50 mM NaCl and mobile phase B containing 250 mM NaCl. The CEX-HPLC and CZE were employed to determine the charge heterogeneity of the molecule. The CEX results showed comparable results to the innovator product with the biosimilar product demonstrating 15% acidic variant, 10% basic variant and 75% main peak, respectively. The CZE and CEX demonstrated similar trending for the charge variants and chromatographic profile. The charge heterogeneity demonstrated no significant differences in the in vitro potency and pharmacokinetic behavior of the biosimilar and Avastin/Bevacizumab used as a reference. Khawli and colleagues separated and characterized the major charge variants (acidic, basic, and main peak fractions) of a recombinant humanized IgG1 via CEX and investigated the effect of the isolated charge variants on the binding affinity and pharmacokinetic behavior in rats (Khawli et al., 2010). No changes in the secondary and tertiary structure were observed (near and far UV region) and pI value of acidic variant, basic variant, and main peak pI were 8.7–8.9, 9.1, and 9.0, respectively as determined by cIEF method. The mass spectrometry data indicated the presence of sialylation, deamidation, reduced disulfide, and glycation in acidic variants and C-terminal lysine clipping was observed on the one or both C-termini of the heavy chain. The in vitro and in vivo data demonstrated no significant impact of the isolated charge variants on the potency, FcRn binding affinity and PK properties of the antibody in healthy rats.

Complete understanding of the modifications in terms of location and their degree to generate acidic and basic species is critical for antibody drug development. C-terminal lysine modification is a very common PTM observed in monoclonal antibodies and the amount of generated lysine variant differs with the different production process (B-cell hybridoma versus CHO). The monoclonal antibodies with presence of lysine (amino acid code: K) on the C-terminal of the heavy chain are more basic than their main species. Van de Bremer and colleagues studied the effect of C-terminal lysine clipping on the complement component activation by engineering antibodies (CD20 and CD38 mAbs) with no charge, a negatively charged side chain or introduction of positively charged 1–3 amino acid residues on the C-terminal of the antibody (K0, K1 and K2 isoforms) (van den Bremer et al., 2015). The isolated isoforms from CEX demonstrated maximal CDC activity in the cells after removal of lysine residues from both the heavy chains with a CPB indicating that C-terminal lysine clipping is critical for antibodies full cytotoxic potential. Along with glycosylation, C-terminal lysine clipping is an important factor that needs to be considered for development of therapeutic antibodies and tuning their effector functions (ADCC and CDC) particularly in expression systems lacking CPB activity. To summarize, data from different studies indicates that effect of charge heterogeneity on the structure, stability, binding kinetics and biological behavior (potency, immunogenicity) in therapeutics monoclonal antibodies varies from molecule to molecule, thus emphasizing on the need to characterize the heterogeneity and assuring batch to batch consistency to generate safe and efficacious antibody drugs.

2.2.4 Identification

Identity of the monoclonal antibody as a therapeutic drug should be highly specific and should be determined by the uniqueness of its molecular structure and/or other specific properties (International Conference, 1999). Generally, the identity of the monoclonal antibody is assessed by peptide mapping which is used to perform an extensive characterization of a protein drug, providing primary amino acid sequence confirmation as well as the identification and quantification of PTMs. In therapeutic antibodies, PTMs such as glycosylation, deamidation, oxidation and isomerization can negatively affect the antibody binding to the FcγRs, resulting in conformational changes and increase in immunogenicity (Hermeling, Crommelin, Schellekens, & Jiskoot, 2004; Liu et al., 2008; Wang et al., 2011; Wei et al., 2007). Monitoring PTMs, in particular those that affect the safety and efficacy of the product and considered as CQAs, is thus important during drug development, production and storage.

The most common peptide mapping approach adopted for analyzing multiple PTMs is the bottom-up LC-MS approach due to its high detection sensitivity and selectivity of mass spectrometry. In the conventional LC-MS methods, proteins are denatured in guanidium hydrochloride (GuHCl) or urea, reduced with DTT which is then followed by alkylation and/or buffer exchange step. The proteins are treated with proteases such as trypsin followed by separation by RP-HPLC hyphenated to a mass spectrometer. The sample digestion step is a time consuming step which can take 16–18 h at 37°C and could potentially generate artifacts such as asparagine (Asn) deamidation, methionine (Met) oxidation and N-terminal glutamine cyclization. Wang et al. developed an ultra-fast bottom-up LC-MS method (uLC-MS) with 5 min tryptic digestion to analyze multiple PTMs (Wang et al., 2016). The uLC-MS eliminates the time consuming alkylation, buffer exchange and UV measurement step present in the conventional method. The sequence coverage of the peptides for glycan profiling analyzed by uLC-MS was 96.5% compared to 93.4% with the conventional method. In addition, the glycan profile obtained from uLC-MS method was similar to the profile generated from the instant 2-AB/HILIC UPLC method. In order to identify specific oxidation, deamidation and isomerization sites, the mAb was stored at different temperatures for 12 months. Compared to the conventional LC-MS method, the percentages of the deamidation, isomerization and oxidation was lower in the uLC-MS. The higher percentages of the PTMs in conventional method (120-fold increase) were attributed to the longer digestion time (18 h) resulting in generation of the artifacts in Asn deamidation, Met oxidation and aspartic acid (Asp) isomerization. The uLC-MS method can be a useful tool to quickly analyze samples at any stage of drug development, greatly reducing the time and labor cost of conventional LC-MS method.

Peptide mapping is a product specific test and is commonly used as a lot release identity test with acceptance criteria "comparable to reference standard." In order to compare the primary structure of the biosimilar CT-P10 and innovator EU- and US-Rituximab, Lee and colleagues used peptide mapping (LC/MS) method (Lee et al., 2018). The samples were reduced and alkylated followed by tryspin digestion at 37°C and peptides separation by reversed phase-HPLC (RP-HPLC) using a C18 column. The absorbance data generated at 214 nm and 280 nm revealed a highly similar peak profile among all the samples without missing or additional new peaks with comparative retention time of each peak, suggesting identical primary sequence of the innovator and the biosimilar products. In the same study, the PTMs were identified and quantitated using LC-MS peptide mapping analysis in which the samples were reduced, alkylated and subjected to trypsin (37°C for 2 h) and asparaginase (37°C for 16 h) digestion. The resulting peptides were separated by RP-UPLC using a C18 column and detected by electrospray ionization mass spectrometry (AB Sciex Triple TOF 5600). The results showed highly similar levels of deamidation and oxidation in both the heavy chain and LC of all the 3 products (<4%), thus demonstrating similarity in the post translational modifications of the CT-P10 with the innovator product.

2.2.5 Higher order structure

Proteins are complex three dimensional structures that change their conformation with the changing environment around them. The folding and shape of the protein directly impacts the functionality of the therapeutic product. Determining or monitoring the conformational changes in monoclonal antibodies and eventually understanding their HOS increases the knowledge about the product and offers in depth understanding about the impact of the process changes on the product quality which may lead to improving both the process and product performance. According to ICH Q5E guideline, changes to the manufacturing process should be followed by determining the product's higher order structure (secondary, tertiary, and quaternary structure) in order to ensure that product conformation is maintained (ICH harmonised tripartite guideline, 2004). Presently, the assays performed to determine the higher order structure of monoclonal antibodies are not generally used for lot release testing, stability testing but they are used for early and late stage product characterization and analytical similarity exercise.

A range of analytical techniques that are routinely employed for analyzing higher order structure quality attribute includes but is not limited to circular dichroism (CD), fourier transform infrared spectroscopy (FTIR), intrinsic fluorescence, extrinsic fluorescence and differential scanning calorimetry (DSC). CD works on the principle of differential absorption (or unequal absorption) of the left and right circularly polarized light and is used for secondary and tertiary structure analysis by far-UV and near-UV CD methods depending on the wavelength range used. Far-UV CD (190–250 nm) is directly linked to secondary structure of proteins due to asymmetrical packing of the achiral peptide bonds whereas the near UV-CD (250–350 nm) is correlated to the specific interactions within the aromatic amino acid side chains and/or their interactions with the peptide group(s) (Tetin et al., 2003). At this wavelength range (250–350 nm), the chromophore is the peptide bond. α-helix, β-sheet and random coils together give rise to a characteristic CD spectrum. The near-UV CD signal is influenced by the microenvironment around the aromatic amino acid residues (phenylalanine, tryptophan and tyrosine) and disulfide bonds. CD data is generally reported in units of ΔE ($E=$ ellipticity) which is the difference in absorbance of E_R and E_L by an asymmetric molecule, or in degrees ellipticity, which is defined as the angle whose tangent is the ratio of the minor to the major axis of the ellipse. $[\theta]$, the molar ellipticity in deg. cm^2 dmol^{-1} $= 3298\Delta E$.

Ejima et al. studied the effect of acidic pH (low pH, pH 2.7 and 3.5) stress on the conformation of two human monoclonal antibodies (hIgGA-4 and hIgGB-4) and a mouse monoclonal antibody (mIgG1) using circular dichroism (Ejima et al., 2007). The near-UV CD spectrum for low pH hIgGA-4 samples was characterized by a positive peak at 292 nm (attributable to tryptophan) and several negative peaks between 260 and 290 nm and was comparable to native protein spectrum of the same antibody at pH 6.0 in terms of presence of aromatic signals. However, compared to native spectra at pH 6.0, a slight upward shift of the spectra was observed at pH 3.5 and significant upward shift was observed at pH 2.7 indicating conformational changes in acid stressed samples which could be due to configuration changes in disulfide bonds in low pH conditions. The far-UV CD spectrum of hIgG A-4 at pH 3.5 was indistinguishable from that at pH 6.0 but the negative CD intensity peak at pH 2.7 is enhanced suggesting the alteration of the secondary structure and increase in the β-sheet structure which is in alignment with the near-UV CD data for pH 2.7. The CD spectrum results generated for other antibodies—hIgGB-4 (pH 3.5) and mIgG1 (pH 2.7)—were comparable and similar in trend to hIgGA-4. The CD analysis from the study demonstrates that aromatic acids conformation remains unchanged in low pH solution, indicating the tertiary structure is present. Furthermore, IgG secondary structure was also retained at low pH conditions.

Fourier transform infrared (FTIR) spectroscopy is another characterization method that is used for determining the secondary structure of monoclonal antibodies (orthogonal to far-UV CD method) and provides qualitative and quantitative information about the secondary structure of the proteins such as α-helix, β-sheet (intramolecular and intermolecular), β turns and side chain vibrations. The FTIR works on the principle of absorption of infrared light of certain wavelengths by the molecule resulting in protein molecule to vibrate. In the case of monoclonal antibodies (or proteins in general), the amide-I bond ($1600-1700\ cm^{-1}$) is the most intense absorption band followed by amide-II bond ($1550\ cm^{-1}$). The individual amide-I bond performs a certain vibration (C = O stretching vibration—70%–80% and stretching of the C-N bond—10%–20%) while the amide-II bond is governed by in-plane N-H bending (40%–60%), C-N (18%–40%) and C-C (10%) stretching vibrations. The entire signal of all the amide bonds in monoclonal antibodies (or protein) produces the amide-I absorption band. The form and position of this amide-I bond is very specific concerning the secondary structure of the protein and is influenced by the changes in pH, temperature and salt concentration.

Lin and colleagues performed comparability studies after the samples were stressed and assessed the use of CD and FTIR compared with other indirect methods of structural characterization such as SEC and IEX that are sensitive and quantitative to detect subtle variations in mAbs (Lin, Glover, & Sreedhara, 2015). The IgG sample was degraded by exposure to following stress conditions: acidic (pH 3.2 for 1 day at 40°C), basic (pH 8.5 for 3 days at 40°C), light (1.2 million lux-hours – visible light), heat (40°C for 42 days), and oxidation (0.5% tBHP, 24 h, ambient). No significant differences were observed in the CD spectra of the control versus the stressed samples, whereas the FTIR spectra for all the samples showed amide-I and amide-II bands at 1640/1690 and $1550\ cm^{-1}$, respectively with barely discernable signal differences. On the other hand, chromatographic data generated using SEC showed loss in the monomer peak in the stressed samples compared to control which was primarily due to increase in the percentage of the high molecular weight species/aggregates and increase in the fragments. Additionally, the IEX results showed percentage main peak loss in stressed samples resulting in an increase in the acidic peaks in the high pH, high temperature and light stressed samples and basic peaks remained constant but the increase in percentage basic peaks was observed in low pH (acidic) stressed samples. The results from this work clearly indicates that CD and FTIR are not sensitive enough to detect the small changes as observed in SEC and IEX method. According to the author(s), this could be because of the amount of the unfolded protein is so small (or large) which is insignificant compared to the remaining folded protein of the native protein and such a small change can only be detected based on the retention time shift of the main peak in chromatographic methods and minimally affect the protein backbone configuration or the aromatic amino acid residue orientation.

Intrinsic fluorescence is another excellent comparative tool and is complementary to CD and FTIR. Intrinsic fluorescence is due to aromatic amino acids, mainly hydrophobic tryptophan (Trp), and shifts in emission spectra of tryptophan is due to conformational changes, denaturation or binding all of which can affect the local environment surrounding the indole ring in tryptophan (Chen & Barkley, 1998). This spectroscopic technique monitors the environment of the tryptophan residues by measuring intrinsic fluorescence as a function of temperature and henceforth, provides information on the changes in the protein structure. Garidel and colleagues investigated the effect of thermal and chemical stressed conditions on the emission spectra of the tryptophan using intrinsic tryptophan fluorescence spectroscopy in IgG monoclonal antibody (Garidel, Hegyi, Bassarab, Weichel, & Rapid, 2008). The monoclonal antibody used in the study has 16 tryptophan residues distributed over all IgG domains. Thermal denaturation (protein sample heated above 80°C) of the antibody resulted in λ_{max} shift from 330 to 337 nm whereas chemical induced denaturation (6 M GdnHCl) shifted the emission maximum to 352 nm compared to the native protein and showed increase in fluorescence intensity by 1.5-fold. The presence of GdnHCl in Trp solution had no influence on the Trp emission maximum wavelength indicating changes in the emission maximum wavelength is due to the change of the polarity in the vicinity of the Trp of the unfolded protein and not due to direct interaction of the GdnHCl with Trp amino acid residues. In addition, the results also suggest that GdnHCl treatment caused loss of structural integrity and exposure of Trp residues to polar phase and thus increase in the fluorescence intensity and λ_{max} shift (referred to as red shift which is commonly observed with increase in the hydrophilicity). On the other hand, in the case of thermal induced denaturation, insoluble aggregates were formed which prevented the exposure of all the Trps to the polar water phase resulting in λ_{max} of 337 nm and reduced fluorescence intensity compared to GdnHCl treated sample. The extrinsic fluorescence analysis of mAbs will be discussed in detail in Chapter 6, Higher Order Structure and Free Thiol, of this book.

The thermodynamic stability such as unfolding of monoclonal antibodies is assessed by DSC. DSC is a thermo-analytical technique that measures heat capacity (C_p) as a function of temperature. It works on the principle of measuring the difference in the amount of heat required to increase the temperature between the sample and the reference cell lacking protein as a function of temperature. DSC has been widely used to evaluate the conformational stability and also provide the melting temperature/unfolding temperature (T_m) of the antibody. It has been shown that unfolding of human IgG1 using DSC at pH 5.5 show two transistions with T_m around 70°C and 82°C and with amplitude of first peak much larger than that of the second peak (Zav'yalov & Tishchenko, 1991). Further studies have also reported that Fab and Fc fragments of the same antibody have melting temperature of 70°C for Fab fragments and 66°C and 82°C for the Fc fragment using DSC (Tischenko et al., 1998). Taking this data into account, it can be inferred that the first melting temperature in an intact IgG1 antibody is associated with melting of the CH_2 domain in the Fc fragment and the melting of the Fab fragment while second transition mainly represents the unfolding of the CH_3 domain of the Fc fragment.

Maity et al. studied the effect of arginine hydrochloride on thermal stability and percentage of aggregate formation of four IgG1 monoclonal antibody using DSC and SEC in dark controlled and light exposed samples (Maity et al., 2009). Arginine and its derivatives are used as solvent additive to prevent protein aggregation, enhance protein solubility and prevent protein loss due to formation of inclusion bodies (Arakawa et al., 2007; Ejima, Yumioka, Arakawa, & Tsumoto, 2005). With the exception of one mAb used in the study the rest of them showed increase in the thermal stability of the dark control samples in PBA (phosphate buffer in the presence of ArgHCl) compared to samples dissolved in PBS buffer (phosphate buffer in sodium chloride). The unfolding temperature peak for the Fab fragment increased by 0.3°C, 0.8°C, and 1.1°C suggesting that arginine hydrochloride enhances antibody stability. However, no significant change in the transition temperatures was observed in samples exposed to light in either PBS or PBA buffers. The SEC analysis demonstrated photo-induced monomer loss was reduced by more than 50% in the presence of the arginine containing PBA buffer in all the antibodies, however, no significant differences were observed in dark controlled samples in PBS and PBA buffers. Overall, the data from this work demonstrates that arginine is an effective additive to enhance antibody thermal stability and suppress light induced aggregation.

2.2.5.1 Analytical similarity assessment of higher order structures in monoclonal antibodies

Comparative analysis of the higher order structure for different batches is an essential part of the structural similarity studies and is assessed between the originator and biosimilar product. Lee et al. utilized 12–15 lots of the in-house biosimilar product CT-P10 and the innovator product Rituxan and MabThera for the comparison (Lee et al., 2018). The secondary and tertiary structures were analyzed by FTIR, far-UV CD and near-UV CD while the thermal stability and thermal transition temperatures were assessed using DSC. The FTIR data demonstrated presence of amide-I and amide-II bands at 1635 cm^{-1} and 1550 cm^{-1} frequency, respectively along with presence of three minor bands between 1200 and 1800 cm^{-1} in FTIR spectra. The FTIR spectra in terms of location and position of the bands was comparable in the biosimilar and innovator product. The DSC thermogram for CT-P10, Rituxan and MabThera used two states, three states transitioning model and T_m observed were at 71°C for CH_2 fragment, 76°C for Fab fragment and 83°C for CH_3 fragment. The thermal stability was comparable in terms of thermal transition temperatures. The antibody secondary and tertiary structures were studied using CD spectroscopy. The optical path length of 1.000 and 0.100 cm were used for near-UV CD and far-UV CD respectively. The far-UV CD spectrum of CT-P10, Rituxan and MabThera showed typical antibody spectrum with a mean minimum at 217 nm and mean maximum at 202 nm, suggesting the secondary structure is dominated by β-sheet motif. The near-UV CD spectrum CT-P10, Rituxan and MabThera showed typical shape of the antibody with mean high maximum at 297 nm and multiple small peaks between 255 to 280 nm resulting from the aromatic amino acids and cysteines (or disulfide bonds) in the antibody. The obtained CD spectra for representative lots were comparable for near-UV CD and far-UV CD spectra with no significant changes in the residual molar ellipticity for biosimilar and innovator product. Overall the generated data demonstrated that CT-P10 is structurally similar to the reference product, Rituxan and MabThera from HOS standpoint.

Saleem and colleagues assessed the analytical similarity between ABP 710, a biosimilar to Infliximab (Remicade), and innovator US- and EU-Infliximab itself using 3–6 lots (Saleem et al., 2020). The secondary structure analysis by FTIR exhibited a strong β-sheet around 1639 cm^{-1} together with a β-sheet band around 1689 cm^{-1}. The profiles for both ABP 710 and Infliximab were visually similar and ABP 710 demonstrated >95% spectral similarity relative to Infliximab. The tertiary structural similarity assessment on ABP 710 and Infliximab was performed using near-UV CD. The near-UV CD profiles were visually similar and all the spectra contain signals with peaks from tryptophan (255 to 275 nm), tyrosine (275 to 285 nm) and phenylalanine (285 to 295 nm), superimposed on the broad disulfide signal from 250 to 285 nm. ABP 710 demonstrated >95% spectral similarity relative to Infliximab and tertiary structures for ABP 710 and Infliximab were similar by near-UV CD spectroscopy. The conformation and thermal stability was assessed by DSC. ABP 710 and Infliximab have two similar endothermic transition temperatures which correspond to unfolding of antigen binding site/CH_2 domain of the heavy chain (T_m1) and CH_3 domain of the heavy chain (T_m2). The results show that both transition temperatures T_m1 of ABP 710 and infliximab RP and T_m2 of ABP 710 and infliximab RP were similar. In conclusion, HOS analysis confirms that ABP710 and Infliximab have similar secondary and tertiary structures and conformational stability.

2.3 Glossary

1. Pharmacokinetics (PK) refers to the movement of drugs through the body.
2. Pharmacodynamics (PD) refers to the body's biological response to drugs.
3. Specification: A specification is defined as a list of tests, references to analytical procedures, and appropriate acceptance criteria which are numerical limits, ranges, or other criteria for the tests described. It establishes the set of criteria to which a drug substance, drug product or materials at other stages of its manufacture should conform to be considered acceptable for its intended use (International Conference, 1999).
4. Acceptance Criteria: Numerical limits, ranges, or other suitable measures for acceptance of the results of analytical procedures which the drug substance or drug product or materials at other stages of their manufacture should meet (International Conference, 1999).
5. Antibody Dependent Cell Mediated Cytotoxicity (ADCC): ADCC is the immune mechanism where the Fc bound effector cells such as natural killer cells (NK cells) lyse the target cell which has antigen bound to the antibody (Fab).
6. Complement Dependent Cytotoxicity (CDC): CDC is the effector function of the antibodies where the Fab region of mAb binds to the membrane surface antigens on the target cells, triggering the activation of the complement pathway by bonding protein C1q to the antibodies and formation of membrane attack complex (MAC) and target lysis.
7. Product Related Impurities: These are product variants with properties different from those of the desired product (DS/DP) (International Conference, 1999).

8. Quality Target Product Profile (QTPP): A prospective summary of the quality characteristics of a drug product that ideally will be achieved to ensure the desired quality, taking into account safety and efficacy of the drug product (ICH Harmonised, 2009).

References

Arakawa, T., Ejima, D., Tsumoto, K., Obeyama, N., Tanaka, Y., Kita, Y., & Timasheff, S. N. (2007). Suppression of protein interactions by arginine: a proposed mechanism of the arginine effects. *Biophysical Chemistry, 127*(1–2), 1–8.

Beck, A., Diemer, H., Ayoub, D., Debaene, F., WagnerRousset, E., Carapito, C., ... Sanglier-Cianférani, S. (2013). Analytical characterization of biosimilar antibodies and Fc-fusion proteins. *TrAC Trends in Analytical Chemistry, Volume 48*, 81–95, July–August.

Beck, A., Wagner-Rousset, E., Bussat, M. C., Lokteff, M., Klinguer-Hamour, C., Haeuw, J. F., et al. (2008). Trends in glycosylation, glycoanalysis and glycoengineering of therapeutic antibodies and Fc-fusion proteins. *Current Pharmaceutical Biotechnology, 9*(6), 482–501.

Boyd, P. N., Lines, A. C., & Patel, A. K. (1995). The effect of the removal of sialic acid, galactose and total carbohydrate on the functional activity of Campath-1H. *Molecular Immunology, 32*(17-18), 1311–1318.

Chen, Y., & Barkley, M. D. (1998). Toward understanding tryptophan fluorescence in proteins. *Biochemistry, 37*(28), 9976–9982.

Ejima, D., Tsumoto, K., Fukada, H., Yumioka, R., Nagase, K., Arakawa, T., & Philo, J. S. (2007). Effects of acid exposure on the conformation, stability, and aggregation of monoclonal antibodies. *Proteins, 66*(4), 954–962.

Ejima, D., Yumioka, R., Arakawa, T., & Tsumoto, K. (2005). Arginine as an effective additive in gel permeation chromatography. *Journal of Chromatography. A, 1094*(1–2), 49–55.

Fekete, S., Beck, A., Fekete, J., & Guillarme, D. (2015a). Method development for the separation of monoclonal antibody charge variants in cation exchange chromatography, part I: salt gradient approach. *Journal of Pharmaceutical and Biomedical Analysis, 102*, 33–44.

Fekete, S., Beck, A., Fekete, J., & Guillarme, D. (2015b). Method development for the separation of monoclonal antibody charge variants in cation exchange chromatography, part II: pH gradient approach. *Journal of Pharmaceutical and Biomedical Analysis, 102*, 282–289.

Garidel, P., Hegyi, M., Bassarab, S., Weichel, M., & Rapid, A. (2008). Sensitive and economical assessment of monoclonal antibody conformational stability by intrinsic tryptophan fluorescence spectroscopy. *Biotechnology journal, 3*(9-10), 1201–1211.

Garza, C. E. E., Perdomo-Abúndez, F. C., Padilla-Calderón, J., Uribe-Wiechers, J. M., Pérez, N. O., Flores-Ortiz, L. F., & Medina-Rivero, E. (2013). Analysis of recombinant monoclonal antibodies by capillary zone electrophoresis. *Electrophoresis, 34*(8), 1133–1140.

Goetze, A. M., Liu, Y. D., Zhang, Z., Shah, B., Lee, E., Bondarenko, P. V., & Flynn, G. C. (2011). High-mannose glycans on the Fc region of therapeutic IgG antibodies increase serum clearance in humans. *Glycobiology, 21*(7), 949–959.

Haberger, M., Leiss, M., Heidenreich, A. K., Pester, O., Hafenmair, G., Hook, M., et al. (2016). Rapid characterization of biotherapeutic proteins by size-exclusion chromatography coupled to native mass spectrometry. *mAbs, 8*(2), 331–339.

Hawe, A., Kasper, J. C., Friess, W., & Jiskoot, W. (2009). Structural properties of monoclonal antibody aggregates induced by freeze-thawing and thermal stress. *European Journal of Pharmaceutical Sciences*, *38*(2), 79–87.

He, Y., Isele, C., Hou, W., & Ruesch, M. (2011). Rapid analysis of charge variants of monoclonal antibodies with capillary zone electrophoresis in dynamically coated fused-silica capillary. *Journal of Separation Science*, *34*(5), 548–555.

He, Y., Lacher, N. A., Hou, W., Wang, Q., Isele, C., Starkey, J., & Ruesch, M. (2010). Analysis of identity, charge variants, and disulfide isomers of monoclonal antibodies with capillary zone electrophoresis in an uncoated capillary column. *Analytical Chemistry*, *82*(8), 3222–3230.

He, X. Z., Que, A. H., & Mo, J. J. (2009). Analysis of charge heterogeneities in mAbs using imaged CE. *Electrophoresis*, *30*, 714–722.

Hermeling, S., Crommelin, D. J., Schellekens, H., & Jiskoot, W. (2004). Structure-immunogenicity relationships of therapeutic proteins. *Pharmaceutical Research*, *21*, 897–903.

Hintersteiner, B., Lingg, N., Zhang, P., Woen, S., Hoi, K. M., Stranner, S., et al. (2016). Charge heterogeneity: basic antibody charge variants with increased binding to Fc receptors. *mAbs*, *8*(8), 1548–1560.

Hodoniczky, J., Zheng, Y. Z., & James, D. C. (2005). Control of recombinant monoclonal antibody effector functions by Fc N-glycan remodeling in vitro. *Biotechnology Progress*, *21*(6), 1644–1652.

ICH harmonised tripartite guideline "Comparability of biotechnological/biological products subject to changes in their manufacturing process", Q5E, 2004.

ICH Harmonised Tripartite Guideline Pharmaceutical Development, Q8(R2), August 2009.

International Conference On Harmonisation Of Technical Requirements For Registration Of Pharmaceuticals For Human Use, ICH Harmonised Tripartite Guideline, Specifications: Test Procedures And Acceptance Criteria For Biotechnological/Biological Products Q6B, 1999.

Jefferis, R. (2005). Glycosylation of recombinant antibody therapeutics. *Biotechnology Progress*, *21*(1), 11–16.

Jefferis, R. (2016). Posttranslational modifications and the immunogenicity of biotherapeutics. *Journal of Immunology Research*, *2016* 5358272.

Jung, S. K., Lee, K. H., Jeon, J. W., Lee, J. W., Kwon, B. O., Kim, Y. J., ... Chang, S. J. (2014). Physicochemical characterization of Remsima. *mAbs*, *6*(5), 1163–1177.

Kaur, H. (2021a). Characterization of glycosylation in monoclonal antibodies and its importance in therapeutic antibody development. *Critical Reviews in Biotechnology*, *41* (2), 300–315.

Kaur, H. (2021b). Stability testing in monoclonal antibodies. *Critical Reviews in Biotechnology*, *17*, 1–32.

Khawli, L. A., Goswami, S., Hutchinson, R., Kwong, Z. W., Yang, J., Wang, X., et al. (2010). Charge variants in IgG1: isolation, characterization, in vitro binding properties and pharmacokinetics in rats. *mAbs*, *2*(6), 613–624.

Lee, K. H., Lee, J. 1, Bae, J. S., Kim, Y. J., Kang, H. A., Kim, S. H., ... Chang, S. J. (2018). Analytical similarity assessment of rituximab biosimilar CT-P10 to reference medicinal product. *mAbs*, *10*(3), 380–396.

Li, H., & d'Anjou, M. (2009). Pharmacological significance of glycosylation in therapeutic proteins. *Current Opinion in Biotechnology*, *20*(6), 678–684.

Lin, J. C., Glover, Z. K., & Sreedhara, A. (2015). Assessing the utility of circular dichroism and FTIR spectroscopy in monoclonal-antibody comparability studies. *Journal of Pharmaceutical Sciences*, *Volume 104*(12), 4459–4466.

Liu, D., Ren, D., Huang, H., Dankberg, J., Rosenfeld, R., Cocco, M. J., ... Remmele, R. L., Jr. (2008). Structure and stability changes of human IgG1 Fc as a consequence of methionine oxidation. *Biochemistry*, *47*, 5088–5100.

Liu, L. (2015). Antibody glycosylation and its impact on the pharmacokinetics and pharmacodynamics of monoclonal antibodies and Fc-fusion proteins. *Journal of Pharmaceutical Sciences*, *104*(6), 1866–1884.

López-Morales, C. A., Miranda-Hernández, M. P., Juárez-Bayardo, L. C., Ramírez-Ibáñez, N. D., Romero-Díaz, A. J., Piña-Lara, N., et al. (2015). Physicochemical and biological characterization of a biosimilar trastuzumab. *BioMed Research International*, *2015* 427235.

Lundell, N., & Schreitmüller, T. (1999). Sample preparation for peptide mapping—a pharmaceutical quality-control perspective. *Analytical Biochemistry*, *266*(1), 31–47.

Ma, S. (2005). In A. R. Mire-Sluis (Ed.), *In State of the Art Analytical Methods for the Characterization of Biological Products and Assessment of Comparability* (pp. 49–68). Karger: Basel.

Mack, S., Cruzado-Park, I., Chapman, J., Ratnayake, C., & Vigh, G. (2009). A systematic study in CIEF: defining and optimizing experimental parameters critical to method reproducibility and robustness. *Electrophoresis*, *30*(23), 4049–4058.

Maity, H., O'Dell, C., Srivastava, A., & Goldstein, J. (2009). Effects of arginine on photostability and thermal stability of IgG1 monoclonal antibodies. *Current Pharmaceutical Biotechnology*, *10*(8), 761–766.

Montacir, O., Montacir, H., Springer, A., Hinderlich, S., Mahboudi, F., Saadati, A., & Parr, M. K. (2018). Physicochemical characterization, glycosylation pattern and biosimilarity assessment of the fusion protein etanercept. *The protein journal*, *37*(2), 164–179.

Putnam, W. S., Prabhu, S., Zheng, Y., Subramanyam, M., & Wang, Y. M. (2010). Pharmacokinetic, pharmacodynamic and immunogenicity comparability assessment strategies for monoclonal antibodies. *Trends in Biotechnology*, *28*(10), 509–516.

Ratanji, K. D., Derrick, J. P., Dearman, R. J., & Kimber, I. (2014). Immunogenicity of therapeutic proteins: influence of aggregation. *Journal of immunotoxicology*, *11*(2), 99–109.

Reusch, D., & Tejada, M. L. (2015). Fc glycans of therapeutic antibodies as critical quality attributes. *Glycobiology*, *25*(12), 1325–1334.

Rosenberg, A. S. (2006). Effects of protein aggregates: an immunologic perspective. *The AAPS Journal*, *8*(3), E501–E507.

Salas-Solano, O., Babu, K., Suhr Park, S., Zhang, X., Zhang, L., Sosic, Z., et al. (2011). Intercompany study to evaluate the robustness of capillary isoelectric focusing technology for the analysis of monoclonal antibodies. *Chromatographia*, *73*, 1137–1144.

Salas-Solano, O., Tomlinson, B., Du, S., Parker, M., Strahan, A., & Ma, S. (2006). Optimization and validation of a quantitative capillary electrophoresis sodium dodecyl sulfate method for quality control and stability monitoring of monoclonal antibodies. *Analytical Chemistry*, *78*(18), 6583–6594.

Saleem, R., Cantin, G., Wikström, M., Bolton, G., Kuhns, S., McBride, H. J., et al. (2020). Analytical and functional similarity assessment of ABP 710 a biosimilar to infliximab reference product. *Pharmaceutical Research*, *37*(6), 114.

Scallon, B. J., Tam, S. H., McCarthy, S. G., Cai, A. N., & Raju, T. S. (2007). Higher levels of sialylated Fc glycans in immunoglobulin G molecules can adversely impact functionality. *Molecular Immunology*, *44*(7), 1524–1534.

Seo, N., Polozova, A., Zhang, M., Yates, Z., Cao, S., Li, H., … Liu, J. (2018). Analytical and functional similarity of Amgen biosimilar ABP 215 to bevacizumab. *mAbs*, *10*(4), 678–691.

Shekhawat, R., Shah, C. K., Patel, A., Srinivasan, S., Kapoor, P., Patel, S., et al. (2019). Structural similarity, characterization of poly ethylene glycol linkage and identification of product related variants in biosimilar pegfilgrastim. *PLoS ONE*, *14*(3), e0212622.

Shields, R. L., Lai, J., Keck, R., O'Connell, L. Y., Hong, K., Meng, Y. G., … Presta, L. G. (2002). Lack of fucose on human IgG1 N-linked oligosaccharide improves binding to human Fcgamma RIII and antibody-dependent cellular toxicity. *The Journal of Biological Chemistry*, *277*(30), 26733–26740.

Suba, D., Urbányi, Z., & Salgó, A. (2015). Capillary isoelectric focusing method development and validation for investigation of recombinant therapeutic monoclonal antibody. *Journal of Pharmaceutical and Biomedical Analysis, 114*, 53–61.

Tetin, S. Y., Prendergast, F. G., & Venyaminov, S. Y. (2003). Accuracy of protein secondary structure determination from circular dichroism spectra based on immunoglobulin examples. *Analytical Biochemistry, 321*(2), 183–187.

Tischenko, V. M., Abramov, V. M., & Zav'yalov, V. P. (1998). Investigation of the cooperative structure of Fc fragments from myeloma immunoglobulin G. *Biochemistry, 37* (16), 5576–5581.

van den Bremer, E. T., Beurskens, F. J., Voorhorst, M., Engelberts, P. J., de Jong, R. N., van der Boom, B. G., et al. (2015). Human IgG is produced in a pro-form that requires clipping of C-terminal lysines for maximal complement activation. *mAbs*, 7(4), 672–680.

Vlasak, J., & Ionescu, R. (2011). Fragmentation of Monoclonal Antibodies. *mAbs, 3*(3), 253–263.

Wang, W., Vlasak, J., Li, Y., Pristatsky, P., Fang, Y., Pittman, T., … Ionescu, R. (2011). Impact of methionine oxidation in human IgG1 Fc on serum half-life of monoclonal antibodies. *Molecular Immunology, 48*, 860–866.

Wang, Y., Li, X., Liu, Y. H., Richardson, D., Li, H., Shameem, M., & Yang, X. (2016). Simultaneous monitoring of oxidation, deamidation, isomerization, and glycosylation of monoclonal antibodies by liquid chromatography-mass spectrometry method with ultrafast tryptic digestion. *mAbs, 8*(8), 1477–1486.

Wei, Z., Feng, J., Lin, H. Y., Mullapudi, S., Bishop, E., Tous, G. I., … Schenerman, M. A. (2007). Identification of a single tryptophan residue as critical for binding activity in a humanized monoclonal antibody against respiratory syncytial virus. *Analytical Chemistry, 79*, 2797–2805.

Zav'yalov, V. P., & Tishchenko, V. M. (1991). Mechanisms of generation of antibody diversity as a cause for natural selection of homoiothermal animals in the process of evolution. *Scandinavian Journal of Immunology, 33*(6), 755–762.

Zhao, Y. Y., Wang, N., Liu, W. H., Tao, W. J., Liu, L. L., & Shen, Z. D. (2016). Charge variants of an avastin biosimilar isolation, characterization, in vitro properties and pharmacokinetics in rat. *PLoS ONE, 11*(3), e0151874.

CHAPTER 3

Glycosylation analysis

Christoph Gstöttner[1], Harleen Kaur[2] and Manfred Wuhrer[1]
[1]Leiden University Medical Center, Center for Proteomics and Metabolomics, The Netherlands
[2]Analytical Sciences, Aurobindo Biologics (Unit-XVII) (CurateQ Biologics), Hyderabad, India

3.1 Introduction

Monoclonal antibodies (mAbs) are very complex molecules which show a plethora of different post translational modifications such as glycosylation, oxidation, deamidation, glycation, lysine clipping and several more (Jefferis, 2016; Kaur, 2021). In this chapter, we focus on the biological function and the characterization of glycosylation as a post translational modification. Glycans of mAbs produced in mammalian cell lines (CHO, SP2/0 or NSO) consist of different building blocks. The most abundant ones are galactose (Gal), mannose (Man), N-acetylglucosamine (GlcNAc), fucose (Fuc) and sialic acids, mainly N-acetylneuraminic acid (NeuAc) for CHO cells and N-glycolylneuraminic acid (NeuGc) for SP2/0 and NSO expression systems. Initially, a glucose (Glc)-capped oligosaccharide $Glc_3Man_9GlcNAc_2$ is transferred in the endoplasmatic reticulum (ER) to asparagine in position 297 localized in the CH2 domain of the Fc portion of IgG mAbs (Martin, 1996). After trimming and elongation of these glycans in the ER and Golgi apparatus, the mAbs are secreted into the fermentation broth by the mammalian cell. Each mAb, consists of two heavy chains and contains two Fc N-glycans in total, which contributes only 2%–3% to its molecular weight. In certain cases, additional N-glycosylation sites in other parts of the antibody, such as in the variable domain (Fab), can be observed (Jefferis, 2005; Reusch & Tejada, 2015).

The Fab glycosylation in monoclonal antibodies is believed to not have an impact on the interaction with canonical immunoglobulin receptors (Huang, Biolsi, Bales, & Kuchibhotla, 2006; Jefferis, 2005; Lim, Reed-Bogan, & Harmon, 2008). Therefore in this chapter we want to focus exclusively on the Fc-glycosylation as well as the O-glycosylation, which can occasionally occur in mAbs.

Three different types of N-glycosylation (high mannose, hybrid-type and complex-type glycans) can be found on mAbs. For mAbs expressed

Monoclonal Antibodies
DOI: https://doi.org/10.1016/B978-0-12-822318-5.00002-8
Copyright © 2021 Elsevier Inc and Harleen Kaur. All rights reserved.

in CHO cells high mannose and hybrid structures tend to occur in only minor amounts (Reusch & Tejada, 2015). All these glycans are built from different sugar precursors and share a single core structure containing two GlcNAc residues linked β1,4 and three Man residues. The 2 GlcNAc residues form the chitobiose unit are linked to the asparagine residue and are capped by a β1,4-linked mannose which carries two additional mannoses in α1,3 and α1,6 linkage. In case of complex type glycans these two α-linked Man residues are elongated with GlcNAc in β1,2 linkage. This glycoform is the so called G0 glycoform, which can additionally carry a α1,6 linked fucose on the innermost, asparagine-linked GlcNAc resulting in the G0F glycoform. The G0 or G0F can be further modified by 1 or 2 galactoses β1,4 linked to the outer GlcNAc residues, resulting in G1/G1F and G2/G2F, respectively. The galactoses can be modified by addition of sialic acids, yet these glycoforms are only present at low levels on most mAb biopharmaceuticals (Jefferis, 2006).

Crystallographic structures showed that in case of IgGs Fc N-glycans are partially buried in a cavity (Higel, Seidl, Sörgel, & Friess, 2016; Huber, Deisenhofer, Colman, Matsushima, & Palm, 1976), which forms a horseshoe-like shape, meaning that CH2 and CH3 domains of the antibody are bent outside (Krapp, Mimura, Jefferis, Huber, & Sondermann, 2003). Removal of these glycans affects the thermal stability of mAbs, enhances unfolding under denaturing conditions, increases the susceptibility to proteolytic cleavages with papain and increases aggregation during storage (Zheng, Bantog, & Bayer, 2011).

Due to the fact that the glycans are buried between the two heavy chains and the resulting steric hindrance, mAbs carry mostly diantennary complex type glycans, with higher antennarity hardly observed (Higel et al., 2016). Nevertheless, a plethora of different glycoforms are possible, with certain glycoforms only present in minor abundances. A comprehensive overview concerning the abundance of certain glycoforms is given in a paper of Thomann et al. where they analyzed the glycans of 54 batches of a mAb. The most abundant ones were G0F and G1F followed by G2F. (Thomann, Reckermann, Reusch, Prasser, & Tejada, 2016). Because of the difference in size between different glycoforms, the horseshoe like structure of mAbs are more open or closed allowing the interaction with other proteins (Higel et al., 2016). Therefore different glycoforms can lead to various in vivo effects, due to their interaction with Fc receptors. Of note, even glycoforms that occur at only low levels may have a severe biological impact (Chung et al., 2008; Jefferis, 2005;

Schiestl et al., 2011). For this reason, we want to focus in the following chapter on the biological effects of these glycoforms and finally describe methods for their detection and characterization. Authorities such as the European Medicines Agency (EMA) or the U.S. Food and Drug Administration (FDA) published guidelines, which state that amino sugars, neutral sugars and sialic acids, as well as the antennarity and the site occupancy should be evaluated. Especially the mannose, fucose, galactose and sialic acid content should be determined and the distribution of the main glycoforms closely monitored (Agency, 2008). Therefore glycosylation, but especially certain glycoforms are considered as critical quality attributes (CQAs). This means certain limits need to be defined and respected to guarantee product safety, quality, pharmacokinetic/pharmacodynamics behavior and efficacy. In the following section, these CQAs and their biological function will be discussed. Furthermore, we will describe the pharmacokinetics/pharmacodynamics of certain glycoforms and finally we will show and discuss methods for the characterization of these glycoforms.

3.2 Biological effects of different glycoforms of monoclonal antibodies

MAbs are interacting with several Fc receptors, either upon target binding or without binding to an antigen. Fc receptors are located on the surface of different cells, such as baso- or neutrophil cells, dendritic cells, macrophages or neutral killer cells (Reusch & Tejada, 2015). Fc receptors are diverse in structure and function and comprise the neonatal Fc receptor (FcRn) as well as the family of Fcγ receptors (FcγRs). The FcRn binding of mAbs is not influenced by mAb N-glycosylation (Jefferis, 2012). The FcγRs are the connection between the adaptive and innate immune systems (Hayes, Wormald, Rudd, & Davey, 2016) and can be further grouped according to their biological function into those which activate the immune response of effector cells (FcγRI, FcγRIIa, FcγRIIIa and FcγRIIIb) or inhibit their response (FcγRIIb) (Nimmerjahn & Ravetch, 2008b). Next to FcγRs also the complement factor 1q (C1q) is important for the immune response following antigen binding (Sörman, Zhang, Ding, & Heyman, 2014). Effects of these molecules are complement-dependent cytotoxicity (CDC), antibody-dependent cell-mediated cytotoxicity (ADCC) and antibody-dependent cellular phagocytosis (ADCP) (Ravetch, 1997). Differences in the activation of these pathways can be caused by diverse Fc glycoforms. The effect of different glycoforms to bind and activate, via the various FcγRs and C1q, different immune cells is shown by a plethora of studies (Daeron, 1997; Dekkers et al., 2017; Gerber & Mosser, 2001; Hayes et al., 2016; Higel et al., 2016; Lund et al., 1995; Radaev & Sun, 2002; Ravetch & Bolland, 2001; Ravetch, 1997; Sondermann, Kaiser, & Jacob, 2001).

As mentioned above, next to the main glycoforms G0F, G1F and G2F, which carry a α1,6 linked core fucose, also afucosylated glycoforms are present in mAb samples. It has been shown that these glycoforms have an up to 50 fold higher affinity to the FcγIIIa receptor (Dekkers et al., 2017; Shinkawa et al., 2003). The interaction between the Fc portion and the FcγIIIa receptor is mediated via unique glycan-glycan and glycan-protein interactions. Studies have shown that a glycosylation at Asn162 on the FcγIIIa receptor is important for this interaction (Ferrara et al., 2011; Ferrara, Stuart, Sondermann, Brunker, & Umana, 2006b; Zeck, Pohlentz, Schlothauer, Peter-Katalinić, & Regula, 2011). Furthermore, crystallographic structures of a complex between Fc portion of a mAb and FcγRIIIa have revealed that the fucose on the IgG Fc glycan is in sterical hindrance with the Asn162 glycosylation of the FcγRIIIa and leads to decreased affinity. However, if one of the two heavy chains of the mAb is carrying an afucosylated glycan, the affinity can be restored (Ferrara et al., 2011). Because the FcγIIIa receptor is located on immune cells such as macrophages or neutral killer cells, activation of this receptor leads to ADCC activity. Thus, afucosylated mAbs, with their higher affinity to FcγRIIIa can lead to increased ADCC activity, which has been shown in several in vitro studies (Chung et al., 2012; Ferrara et al., 2006a; Iida et al., 2009; Jefferis, 2007; Li et al., 2017; Okazaki et al., 2004; Peipp et al., 2008; Pereira, Chan, Lin, & Song, 2018; Shields et al., 2002; Umana, Jean-Mairet, Moudry, Amstutz, & Bailey, 1999). Interestingly it has been shown by Shatz et al. that afucosylation on only one of the two glycans can fully restore ADCC activity (Shatz et al., 2013). In certain cases, in particular for oncological indications, a high ADCC activity is wanted to eliminate target cells, such as cancer cells. To the best of our knowledge there are currently three glycoengineered antibodies approved which show low levels of fucosylation and therefore high ADCC response (Pereira et al., 2018). Still, for applications where a high ADCC reaction is unwanted as well as for mAbs, which show off-target binding, the level of afucosylation has to be monitored closely. Next to the ADCC response, ADCP response was also shown to be increased for afucosylated mAbs, in case of $CD16^{+}$ monocytes and M2c macrophages, whereas for

CD16^{-} monocytes and M1 macrophages the ADCP activity of mAbs with either core fucosylation or without the core fucose was the same (Herter et al., 2014). No difference in CDC response between afucosylated and α1,6 core fucosylated mAbs has been found (Chung et al., 2012; Dekkers et al., 2017).

As mentioned before, the main glycans of mAbs produced by CHO cells are G0F, G1F and G2F, with 0,1, or 2 terminal galactoses, respectively. Nevertheless, the effect of galactosylation is not completely understood and difficult to study because of interfering effects of other glycoforms. Certain studies claim no connection between different levels of galactosylation and ADCC response (Boyd, Lines, & Patel, 1995; Hodoniczky, Zheng, & James, 2005; Shinkawa et al., 2003). However, it has been shown that IgGs with higher levels of galactosylation exhibit an increased binding affinity to FcγRII and FcγRIIIa (Houde, Peng, Berkowitz, & Engen, 2010; Ritamo, Cloutier, Valmu, Neron, & Rabina, 2014). This was also confirmed by a study of Dekkers et al. where they showed that decreased fucosylation levels show an increase in the binding affinity to the FcγRIIIA and FcγRIIIB by a factor of 17 whereas in combination with hypergalactosylation the increase in binding affinity is by a factor of 40 (Dekkers et al., 2017). Another study by Thomann at al. compared the ADCC response of 54 batches of a mAb and found associations not only with afucosylation levels, but interestingly also with galactosylation (Thomann et al., 2016). Additionally in this study mAbs were enzymatically hypergalactosylated and a positive effect on the ADCC response could be observed (Thomann et al., 2016). The same was shown in a previous study (Jiang et al., 2011). This study also showed that the ADCC response of glycoengineered antibodies with high levels of afucosylation could not be increased by higher levels of galactosylation. The main response was due to the high levels of afucosylation, which is in line with results of Ritamo et al. (Ritamo et al., 2014). In a recent study, differences in binding affinity to FcγRII and FcγRIII, depending on the different G1F isomers were shown. The G1F isoform with the galactose on the α1,6 mannose arm showed higher affinity than the isoform with the galactose on the α1,3 mannose arm (Aoyama et al., 2019).

High galactosylation is also known to induce in vitro CDC response via binding to C1q (Boyd et al., 1995; Dekkers et al., 2017; Tsuchiya et al., 1989) additionally leading to activation of the complement system (Hodoniczky et al., 2005; Jefferis, 2009; Nimmerjahn, Anthony, & Ravetch, 2007). Also looking at the G1F isomers, a higher affinity to C1q and a higher CDC response was shown for the G1F isomer with the galactose located on the α1,6 mannose arm (Aoyama et al., 2019).

Using SP2/0 or NSO as a production system for mAbs, an additional antennary α1−3 galactose can be observed (Chung et al., 2008; Larsen et al., 1989; Sheeley, Merrill, & Taylor, 1997), whereas this glycan species normally cannot be observed on mAbs produced using CHO cells, due to the lack of the necessary enzyme (Jenkins, Parekh, & James, 1996). While most humans have approximately 1% IgG antibodies against α1−3 galactose some people show additionally anti-α1−3 galactose IgE antibodies in the bloodstream (Galili, 2013), which can lead to severe effects as in the case of Cetuximab (Chung et al., 2008) or for Oralatumab (Van Tine et al., 2019). Cetuximab is produced in SP2/0, whereas Oralatumab is produced in NSO cells (Moroncini et al., 2018). While both mAbs have an Fc and a Fab glycosylation site with α1−3 galactose (Van Tine et al., 2019), most of the α1−3 galactose is located in the Fab portion (Qian et al., 2007). The Fc glycosylation seems to be not accessible for the IgE antibodies (Carlsson et al., 2019; Hilger, Fischer, Wölbing, & Biedermann, 2019), which is also in line with what Lammerts van Bueren et al. found in there study (Lammerts Van Bueren et al., 2011). Therefore if antibodies are produced either in NSO or in SP2/0 cells the abundance of α1−3 galactose-containing glycoforms should be monitored.

Sialic acid can be linked to the terminal galactose of a mAb glycan in the form of NeuAc or NeuGc. CHO cell-produced mAbs carry very low amounts of sialic acids in general (Jefferis, 2006) and even to a way lower extent NeuGc, whereas they can be highly abundant in mAbs produced by Sp2/0 or NPO cells (Raju, 2003; Raju, Briggs, Borge, & Jones, 2000). In several studies NeuAc has been shown to have anti-inflammatory effects (Anthony et al., 2008a; Anthony, Wermeling, Karlsson, & Ravetch, 2008b; Kaneko, Nimmerjahn, & Ravetch, 2006; Nimmerjahn & Ravetch, 2008a; Scallon, Tam, Mccarthy, Cai, & Raju, 2007) which might be linked to its interaction with the type II FcRs, such as DC-SIGN or other lectins (Anthony et al., 2008b) although recent observations indicate that the molecular basis of this anti-inflammatory effect is still poorly understood (Temming et al., 2019). Following the lower affinity to other Fcγ receptors also ADCC might be downregulated. However, ADCC responses have been found not to be influenced by sialic acids (Lin et al., 2015). However in this study only glycans without core fucosylation were investigated. The same effect for sialylated glycans without core fucose was also shown by Li et al. (Li et al., 2017). Looking at glycans with a core fucose, being the most abundant glycoforms in a mAb sample and sialylated species there was a slight reduction in the affinity for the FcγRIIIa receptor, as well as a significantly lower ADCC response. A recent study from Thomann et al. measured the binding to different Fcγ receptors and the ADCC activity using glycoengineered, high and low fucosylated mAbs with sialic acids either linked 2,3 or 2,6 to the terminal galactose (Thomann et al., 2019). Depending on the linkage they found that the 2,3 linked sialic acid showed lower affinity to the Fcγ receptors and also lower ADCC, whereas the 2,6 linked sialic acid showed increased affinity as well as ADCC response compared to the wildtype material. Also it was shown that a higher level of sialylation of galactosylated IgGs, resulted in a higher C1q binding, but not in an increased CDC response (Dekkers et al., 2017). As the amount of sialic acid is normally low in CHO cell produced mAbs, this might have only a minor effect.

In contrast, NeuGc which can be highly abundant in mAbs produced by SP2/0 and NSO and present in minor amounts also in CHO cell produced proteins (Hokke et al., 1990; Noguchi, Mukuria, Suzuki, & Naiki, 1996; Raju et al., 2000; Raju, 2003), should be monitored closely. Also the culture medium was shown to have an influence on the amount of NeuGc in the secreted therapeutic protein (Ghaderi, Taylor, Padler-Karavani, Diaz, & Varki, 2010). Humans are unable to produce NeuGc due to a mutation of the gene for the converting enzyme (Varki, 2007). It is believed that healthy individuals do not show adverse reactions to NeuGc-containing biopharmaceuticals. More and more studies show that most of the humans have anti-NeuGc antibodies most likely due to their diet (Padler-Karavani et al., 2008; Tangvoranuntakul et al., 2003; Zhu & Hurst, 2002) and that they develop them already from an age of 6 months (Taylor et al., 2010). A more recent study has also shown that mAbs carrying overall only one NeuGc molecule on the Fc-glycans were not bound by anti-NeuGc-antibodies, whereas mAbs with two to four NeuGc molecules were bound (Yu et al., 2016). It was shown that mAbs with NeuGc glycans may be cleared fast, but more importantly they may prevent re-administration of a drug containing NeuGc due to a potential high immune response (Durocher & Butler, 2009).

High mannose glycans occur during the production of mAbs in all three cell lines mentioned in this chapter. To test whether they have an impact on the biological response Kanda et al. compared the binding to FcγRIIIa and the ADCC response to complex and hybrid type glycans. It was shown that the binding to FcγRIIIa and ADCC were increased compared to fucosylated complex or hybrid type glycans, but lower than for the afucosylated complex type glycans. They did not observe a difference between M5, M8 and M9 structures. Also the binding to C1q and the CDC response were lower than for mAbs with complex-type glycans (Kanda et al., 2006). This effect was also observed by other studies, where they saw a decrease of up to 50% of CDC response compared to other type of glycans. Furthermore the binding to FcγRIIa and FcγRIIb was also shown to be decreased for high mannose glycans (Yu et al., 2012). However it remains unclear if the lack of fucose is responsible for the increase in binding or the high mannose structure itself.

Antibodies can lack one or two antennae of their Fc N-glycans or contain so-called non glycosylated heavy chains. It has been shown that these glycans do not influence the immunogenicity (Jung, Kang, Kelton, & Georgiou, 2011) but have a strong impact on the biological activity of the antibody. Studies have shown that the affinity of aglycosylated antibodies to FcγRI is by two orders of magnitude lower compared to the wildtype antibody. In addition, the affinity to FcγRII and FcγRIII is completely lost (Jefferis, 2009; Jung et al., 2010; Mimura et al., 2000; Tao & Morrison, 1989). Also the affinity to C1q was dramatically decreased for aglycosylated mAbs (Lund, Takahashi, Pound, Goodall, & Jefferis, 1996). This reduced or completely lost affinity to FcγR as well as to C1q also abrogates ADCC or CDC (Jefferis, 2009; Jung et al., 2010, 2011; Lund et al., 1996; Tao & Morrison, 1989). One study showed the effect of hemi-glycosylated mAbs on FcγR receptors and C1q (Ha et al., 2011). As in the case of completely aglycosylated mAbs, the hemiglycosylated ones also show decreased binding affinity to all Fcγ receptors, decreased ADCC as well as a slightly decreased affinity to C1q. Shatz et al. produced antibodies with defined Fc pairings using knob into hole technologies. They could show that in a fully glycosylated mAb, one fucose is enough to restore the ADCC activity compared to a wild-type mAb. It was also shown that a hemi-glycosylated and hemi-glycosylated-hemi-afucosylated mAb (meaning only one heavy chain carries a glycan without core fucose) had lower potency compared to wildtype. Although a hemi-glycosylated-hemi-afucosylated mAb can partially restore ADCC, this effect however is much stronger for a fully glycosylated mAb (Shatz et al., 2013).

3.3 Pharmacokinetic effects of different glycoforms of monoclonal antibodies

There are in principle three different molecular interactions that can influence the pharmacokinetics (PK) of an antibody. The antigen binding (as well as the off target binding), binding by glycan receptors and the FcRn mediated recycling (Liu, 2018). In this chapter we want to focus on the influence of glycans on PK and mAb interaction with different receptors.

Starting with the pharmacokinetics of aglycosylated mAbs the results are not quite clear. Several studies have shown that removal of the glycans lead to a conformational change and instead of the horseshoe like structure a closed one was observed (Feige et al., 2009; Krapp et al., 2003). Others however showed under more physiological conditions that removal of the glycans resulted in a more open structure (Alsenaidy et al., 2013, 2014; Borrok, Jung, Kang, Monzingo, & Georgiou, 2012). Additionally this change of conformation can lead to higher levels of aggregation. This has been shown in several studies (Hristodorov et al., 2013; Kayser et al., 2011). Nevertheless, it is still unclear whether deglycosylated or aglycosylated mAbs have a different PK behavior. One study was comparing the PK of recombinantly produced mAbs, as well as mouse monoclonal antibodies with their corresponding mutants where the asparagine at position 297 was exchanged by an alanine. In this study a lower half-life for the aglycosylated mAbs could be observed (Wawrzynczak, Cumber, Parnell, Jones, & Winter, 1992; Wawrzynczak, Parnell, Cumber, Jones, & Winter, 1989). Other studies using glycosylation depleted antibodies as well as mutated versions of IgG showed that the half-life is independent of the glycosylation and is the same for glycan-depleted/mutated antibodies and their wildtype counterparts (Jung et al., 2011; Kim, Tsen, Ghetie, & Sally Ward, 1994; Liu et al., 2011; Nose & Wigzell, 1983). This was also supported by a phase I trial of an aglycosylated IgG1, which showed a half-life comparable to a wildtype IgG1 (Clarke, Smith, Gebbie, Sweeney, & Olszewski, 2009). Looking at the FcRn mediated recycling it has been shown that glycosylated and aglycosylated mAbs have the same affinity to the receptor which would also suggest that glycosylated and aglycosylated IgG should have comparable half-lifes (Kim et al., 1994; Liu et al., 2011).

For assessing the PK behavior of mAbs with high mannose glycans several studies have been performed showing that these glycoforms are cleared relatively faster compared to other glycosylation types (Goetze et al., 2011; Kanda et al., 2006; Liu et al., 2011; Yu et al., 2012). However, the mechanism or receptors responsible for removal of these mAb glycoforms are not well understood. For high mannose it is assumed that the mannose receptor (ManR) is responsible for removal (Alessandri et al., 2012; Goetze et al., 2011; Yu et al., 2012). Structural studies show that high mannose glycoforms might be accessible by this receptor and are cleared fast from the plasma (Goetze et al., 2011; Kanda et al., 2006; Liu et al., 2011; Wright & Morrison, 1994; Wright et al., 2000; Yu et al., 2012). Furthermore Kanda et al. showed that different high mannose glycoforms (Man5, Man8 or Man9) have the same affinity to FcRn. Of note, their half-life of 4.6 days was way shorter than for mAbs with hybrid type glycans (8.5 days) and complex type glycans (14 days) (Kanda et al., 2006). Other studies (Liu et al., 2011; Yu et al., 2012) including clinical studies observed also a faster clearance of high mannose glycoforms (Chen, Liu, & Flynn, 2008; Goetze et al., 2011). While Chen et al. reported that the faster clearance came from in vivo glycan cleavage, Goetze et al. hypothesized that high mannose glycoforms are cleared faster due to their interaction with the ManR. Next to this, the distribution of mAbs, carrying high mannose glycoforms was investigated and an accumulation in the liver could be observed, which may be an additional hint for the binding to the ManR which is highly abundant on liver cell surface (Wright et al., 2000).

Another important receptor which is abundant on liver cells is the asialoglycoprotein receptor (ASGPR). This receptor binds proteins without sialic acids, so with terminal galactose, N-acetylgalactosamine or N-acetylglucosamine and thereby removing them from the circulation (Spiess, 1990). For mAbs the literature is ambiguous. Some studies found no difference between sialylated and non sialylated mAbs and state that the PK is similar for both variants (Naso, Tam, Scallon, & Raju, 2010; Scallon et al., 2007; Wright & Morrison, 1998). However recently Bas et al. showed that mAbs with hypersialylation of Fc glycans exhibit lower clearance from the circulation compared to wildtype glycoforms (Bas et al., 2019). Regarding the PK several questions remain open, especially regarding the mechanism of clearance of certain glycoforms. However more and more studies are trying to address this gaps.

The biological effect as well as the pharmacokinetic effect of Fc glycans is summarized in Table 3.1.

Table 3.1 Impact of Fc glycans on safety/immunogenicity, biologic activity/efficacy and clearance (PK/PD).

Glycan species	Safety/ Immunogenicity	Biologic activity/Efficacy	Clearance (PK/PD)
Galactose	unknown	+	unknown
alpha1,3-Galactose	—	unknown	unknown
Fucose	(-)	++	unknown
Mannose (high-mannose)	unknown	+	—
NANA	unknown	(-)	+
NGNA	—	(-)	+
Non-glycosylated heavy chain	unknown	—	(-)

+ moderate positive impact, - moderate negative impact, ++ high positive impact, — high negative impact.
Source: Adapted from Reusch & Tejada (2015) Fc glycans of therapeutic antibodies as critical quality attributes. Glycobiology, 25, 1325–1334.

3.4 Characterization of monoclonal antibody glycosylation

Analyzing glycans can be very challenging. One reason for this is that each mAb sample consists of a plethora of different glycoforms with very different relative abundances. A high sensitivity and dynamic range is needed to detect also lower abundant species, which might have a significant biological impact. There are many different approaches to analyze the glycosylation of mAbs. They can be distinguished by the level of analysis.

Glycans can be analyzed either

1. on the intact level
2. on the middle up/down level, either after cleavage with enzymes such as IdeS or SpeB, or after reduction of the disulfide bonds
3. on the bottom-up level (glycopeptides) after digestion with enzymes such as trypsin or GluC or
4. as released glycans after release with enzymes such as PNGase F.

All mentioned approaches come with some advantages/disadvantages and can either be used in combination with mass spectrometry (MS), electrospray ionization, or matrix-assisted laser desorption ionization (MALDI) or as standalone techniques.

In the following we want to show the most common approaches used for the characterization of glycans in the biopharmaceutical industry. As mentioned before one possibility is the analysis of mAbs on the intact level. Most separation techniques, which are able to separate mAb proteoforms on the intact level are hardly compatible with mass spectrometry, such as hydrophobic interaction chromatography or ion exchange chromatography (IEX). However, in order to achieve insights into modifications including glycosylation, efforts are increasingly made to hyphenate these and other techniques with MS. Examples are IEX-MS (Füssl et al., 2019), size exclusion chromatography (SEC)-MS (Brady, Valliere-Douglass, Martinez, & Balland, 2008; Haberger et al., 2016) or sheathless capillary electrophoresis (CE)-MS (Haselberg et al., 2018). These techniques allow not only study of the mAb molecule and its macroheterogeneity, but also enable the assignment of the major glycoforms and the detection of mAb variants missing, for example, the glycosylation on one heavy chain. Additionally the pairing of the glycans on the two Fc parts can be deduced.

Using endoproteases to cleave the mAb in the hinge region allows to analyze mAbs on the middle up/down level. This results in higher sensitivity, also allowing detection of lower abundant glycoforms. In addition, a better separation between the different glycoforms is possible. The most common technique is reversed phase (RP)-high performance liquid chromatography (HPLC)-MS (An, Zhang, Mueller, Shameem, & Chen, 2014; Wagner-Rousset et al., 2008) but also CE-MS (Gstöttner et al., 2020; Haselberg et al., 2018) or hydrophilic interaction liquid chromatography (HILIC)-MS (D'atri, Fekete, Beck, Lauber, & Guillarme, 2017). Another method can be employed to study the glycans on the heavy chain after mAb reduction (Wada, Matsui, & Kawasaki, 2019; Xie et al., 2010).

However, to achieve higher sensitivity, proteins are more regularly digested with proteases resulting in glycopeptides or the N-glycans are enzymatically released using PNGase F. The resulting glycopeptides and released glycans can be analyzed using separation techniques and non-mass spectrometric detection, such as pulsed amperometric detection or fluorescence detection after labeling the glycans with a fluorophore. Next to that there are also a plethora of methods, where either an upfront separation is hyphenated with mass spectrometry, or where mass spectrometry is used as a standalone technique (Zhang et al., 2016). The benefit of "separation only" methods for biopharmaceuticals is their robustness, sensitivity, precision and accuracy. It was also shown that mass spectrometry can lead to biased results due to in-source decay of the glycans during ionization, which needs careful monitoring of MS settings, system suitability tests and stabilization mechanisms such as derivatization of sialic acids (Reiding, Blank, Kuijper, Deelder, & Wuhrer, 2014; Reusch et al.,

2015a). Therefore for release of biopharmaceuticals in a quality control (QC) environment preferably separation-only techniques are used, whereas for the de novo characterization of glycosylation hyphenation to MS is often chosen due to the insights into the structural characteristics provided by the MS dimension.

Separation-only techniques can be further classified by their separation mechanism, either HPLC, for example, HILIC (Fig. 3.1) (Anumula & Dhume, 1998; Melmer et al., 2010) or RP-HPLC for glycoengineered antibodies (Chen & Flynn, 2007) or CE (Kamoda et al., 2004; Kamoda, Ishikawa, & Kakehi, 2006).

For these techniques the glycans are released and labeled with fluorophores by reductive amidation, to allow fluorescence detection. Common labels are procainamide (ProA), 2-aminobenzoic acid (2-AA), 2-aminobenzamide (2-AB), 8-aminopyrene-1,3,6-trisulfonate (APTS), 2-aminopyridine (PA) and 2-aminonaphthalene trisulfonic acid (ANTS)

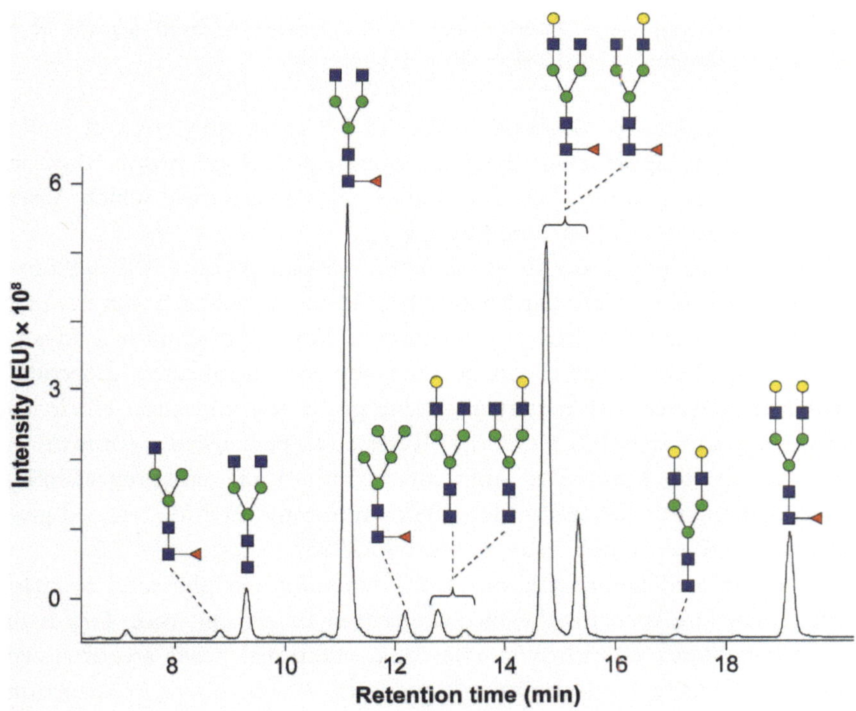

Figure 3.1 2-Aminobenzamide (2-AB) labeled released N-glycans of a mAb sample analyzed by hydrophilic interaction liquid chromatography (HILIC) with fluorescence detection.

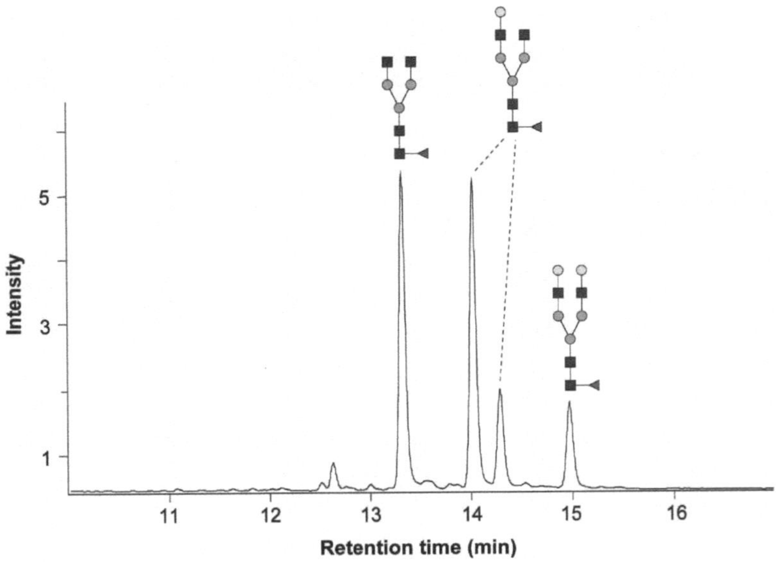

Figure 3.2 8-Aminopyrene-1,3,6-trisulfonate (APTS) labeled released glycans of a mAb sample separated using capillary electrophoresis (CE).

(Ruhaak et al., 2010). Whereas 2-AB and PA are mainly used in liquid chromatography approaches, 2-AA as well as APTS are mainly used in capillary electrophoresis. Fig. 3.2 shows released glycans which were labeled with APTS and separated by CE.

Another recently introduced label for released glycans is *Rapi*Fluor-MS, which enables a very fast labeling procedure in around 5 min because of its highly reactive *N*-hydroxysuccinimide group. An additional advantage of this label is, that it can be used also for fluorescence detection. *Rapi*Fluor increases with its tertiary amine group the ionization efficiency and thereby increases MS sensitivity. Keser et al. performed a comparison with respect to sensitivity, limit of quantification and repeatability between *Rapi*Fluor-MS and other labels commonly used for released glycan analysis (Keser, Pavić, Lauc, & Gornik, 2018).

As mentioned before CE or HPLC techniques hyphenated to mass spectrometry are used for the characterization of glycans, too. This is of special benefit for unknown samples or if additional peaks are observed during a HPLC or CE separation. New labels which allow UV detection and increase the ionization efficiency in positive mode MS were recently described (Hilliard et al., 2017). With this method a HILIC separation can directly be coupled to MS and glycoforms can be assigned with a higher level of confidence and so unknown glycoforms can be characterized. Several articles describe the combination of CE with MS, using either a traditional CE device and APTS as a label (Gennaro & Salas-Solano, 2008) or a microfluidic device in combination with, for example, tandem mass tags (TMT) label (Khatri et al., 2017). Also the analysis of glycopeptides can be either performed directly with HPLC or CE hyphenated to MS, without the need of a labeling procedure (De Leoz et al., 2020).

Next to separation only techniques standalone mass spectrometric approaches based on MALDI are used for the characterization of glycans. The benefit of these techniques is that samples can be measured in a high-throughput manner. However sialic acids need to be derivatized in order to avoid a bias in sialylated glycans through in-source decay (Reiding et al., 2014). A comprehensive oversight concerning the separation- and MS-based methods for glycan analysis can be found in (Reusch et al., 2015a, 2015b).

Recently an interlaboratory study, analyzing the National Institute of Standards and Technology (NIST) antibody glycosylation on different levels, showed that a high variation of results between labs was observed (De Leoz et al., 2020). For this reason it is recommended to use at least two orthogonal methods for glycosylation analysis and adequate system suitability tests. For QC release purposes, with only a limited amount of samples, the 2-AB HILIC separation method is recommended because it shows high robustness. However for glycoengineered antibodies where the afucosylation level is important to monitor also a RP-HPLC approach should be considered. In addition to these HPLC based approaches a CE separation after APTS labeling of glycans might be recommended. In contrast to QC release during clone selection and at-line measurements during mAb production a huge amount of samples have to be processed in a high-throughput manner. In this case multiplexed CE on a DNA analyzer can be applied (Reusch et al., 2014) (Fig. 3.3).

More and more approaches also try to not only characterize the glycosylation of an antibody but also to establish a structure function relationship between the glycans and their receptors. This can either be done by surface plasmon resonance (SPR) but also by affinity liquid chromatography (LC) (Dashivets et al., 2015; Lippold et al., 2019a; Lippold, Nicolardi, Wuhrer, & Falck, 2019b; Thomann et al., 2015). Another trend is to apply a multiple attribute monitoring approach for quality control, which is based on a mass spectrometric measurement of digested mAb samples. This approach allows to quantify different CQAs in only one experiment and would be able to reduce number of measurements, amount of

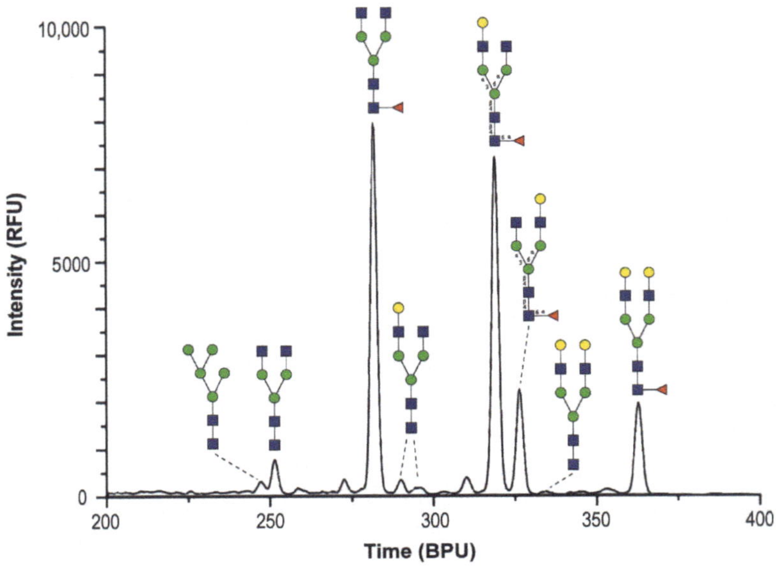

Figure 3.3 8-Aminopyrene-1,3,6-trisulfonate (APTS) labeled released glycans of a mAb sample analyzed by capillary electrophoresis (CE) using a DNA analyzer.

platforms and save thereby time and money (Rogers et al., 2015; Rogstad et al., 2019).

3.5 O-glycosylation of monoclonal antibodies

Another major class of glycosylation that occurs in eukaryotes is *O*-linked glycosylation and involves the transfer of the individual monosaccharide units to the alcoholic side (-OH group) of the serine (Ser) and threonine (Thr) residues (Kaur, 2021). In comparison to *N*-linked glycosylation, the factors regulating *O*-linked glycosylation in mammals is less well understood and less predictable. Unlike *N*-glycosylation which always begins in endoplasmic reticulum (ER) in mammals, the *O*-linked glycosylation synthesis can either start in the endoplasmic reticulum for *O*-fucosylation and *O*-mannosylation or in Golgi apparatus for mucin type *O*-glycosylation (Hülsmeier, Welti, & Hennet, 2011). *O*-glycan are synthesized by several polypeptide N-acetylgalactosamine (GalNAc) transferases, which are connecting GalNAc mainly to Ser or Thr residues.

O-glycosylation is of utmost importance for determining and maintaining the protein structure and protein stability but also for the biological functions, such as immune response and cell signaling (Van Den Steen, Rudd, Dwek, & Opdenakker, 1998). Considering the influence of *O*-linked glycosylation on protein structure and functions, oligosaccharide profiling of the released *O*-glycans is one of the critical tools for early stage monitoring and physicochemical characterization of a monoclonal antibody or a fusion protein. Unfortunately, to our knowledge at present there is no enzyme that can entirely remove *O*-glycans from the protein which led to utilization of chemical based methods such as β-elimination and hydrazinolysis to cleave *O*-glycans from the protein. β-elimination method involves the use of high pH conditions such as use of ammonia or its derivative to release the carbohydrate chain followed by chromophore or fluorophore labeling at the reducing agent. On the other hand, hydrazinolysis initiates by a reaction of anhydrous hydrazine to release the *O*-glycans which is followed by re-N-acetylation step involving addition of acetic anhydride which is then required for attachment of fluorophore or chromophore. Many commercial *O*-glycan hydrazinolysis kits are available from different companies for release of *O*-glycans from proteins (Ludger; Prozyme).

Turyan et al. investigated the efficiency of the ammonia based β-elimination method and hydrazinolysis deglycosylation procedure for quantitative release of the *O*-glycans in fusion proteins (Turyan, Hronowski, Sosic, & Lyubarskaya, 2014). The released *O*-glycans in the study were labeled with 2-aminobenzamide dye followed by HILIC analysis. The chromatographic results showed significant differences in deglycosylation activity between the two methods. The release of *O*-glycans was shown to be 20 to 30 times more efficient with hydrazinolysis than with ammonia based β-elimination. In order to identify the glycan structures, the collected HILIC peak fractions were subjected to Linear Ion Trap—Fourier Transform (LTQ FT) Ultra Hybrid mass spectrometer analysis. FT-MS allowed the assignment of the precursor ions which were further analyzed by LTQ MS/MS. The mass spectrometry data identified several *O*-glycan structures such as mucin-type *O*-glycans (addition of an N-acetylgalactosamine residue (GalNAc) to the hydroxyl group of either Ser or Thr), PentPent-Hex, HexGlcUdehydratedHexPent and GlcU $(Hex)_2$Pent with mass accuracy below 1 ppm.

Enbrel is a dimeric fusion protein built of the tumor necrosis factor receptor (TNFR2, 75 kDa) which is linked to a human IgG Fc portion (Hinge region, CH2 domain and CH3 domain). Enbrel acts as a natural antagonist for TNF and binds it with high affinity. Thereby it prevents the interaction between TNF its receptor and decreases its role in disorders involving inflammation such as rheumatoid arthritis. Cho and colleagues performed comprehensive characterization of *O*-glycans as part of the analytical similarity assessment to evaluate degree of similarity between an in house biosimilar product – SB4 with its reference product Enbrel (EU-sourced) (Cho et al., 2016). The orela reagent was employed to release the *O*-glycan from the sample followed by 2-AB labeling and separation using Acquity UPLC BEH glycan column. The HILIC analysis demonstrated the presence of monosialylated and disialylated *O*-glycan species on the reference product and SB4. In addition, low intensity of *N*-glycan peaks were also observed leading to a conclusion that the *O*-glycan molar concentration is higher than that of *N*-glycan in both SB4 and reference product. For calculating the *O*-glycan occupancy the intact protein mass profile and peak area using reduced, desialylated and *N*-deglycosylated samples were used. The average *O*-glycan occupancy was around 9.3 *O*-glycan units per chain compared to 8.4 *O*-glycan units per chain for SB4. The slightly lower *O*-glycan occupancy in SB4 relative to the reference product was considered to be non-significant and do not negatively influence the biological activities of the biosimilar product.

3.6 Conclusion

Next to *N*-glycosylation certain antibodies can carry *O*-glycosylation, especially if they are expressed in combination with another protein as fusion proteins. Various types of glycosylation of biopharmaceuticals are considered as critical quality attributes due to their in vivo effect. For this reason mAb glycans need to be characterized to guarantee product efficacy and safety. Glycosylation of antibodies can be characterized on different levels (intact, middle-up, peptide or released glycan level). Whereas for the characterization of the glycosylation on the intact and middle-up no additional sample preparation is necessary, glycopeptides and released glycans might either be derivatized to stabilize the sialic acids or labeled to be able to measure them with fluorescence detection to obtain higher sensitivity.

The method to be used has to be carefully considered. It is recommended not to use only one method. It depends also if the method is used as a release method, which would make a validation mandatory. In this case a nonmass spectrometry-based method might be a better choice.

References

Agency, E.M. 2008. Guideline on development, production, characterisation and specifications for monoclonal antibodies and related products https://www.ema.europa.eu/en/documents/scientific-guideline/guideline-development-production-characterisation-specifications-monoclonal-antibodies-related_en.pdf.

Alessandri, L., Ouellette, D., Acquah, A., Rieser, M., Leblond, D., Saltarelli, M., ... Correia, I. (2012). Increased serum clearance of oligomannose species present on a human IgG1 molecule. *MAbs*, *4*, 509–520.

Alsenaidy, M. A., Kim, J. H., Majumdar, R., Weis, D. D., Joshi, S. B., Tolbert, T. J., ... Volkin, D. B. (2013). High-throughput biophysical analysis and data visualization of conformational stability of an IgG1 monoclonal antibody after deglycosylation. *Journal of Pharmaceutical Sciences*, *102*, 3942–3956.

Alsenaidy, M. A., Okbazghi, S. Z., Kim, J. H., Joshi, S. B., Middaugh, C. R., Tolbert, T. J., & Volkin, D. B. (2014). Physical stability comparisons of IgG1-Fc variants: effects of N-glycosylation site occupancy and Asp/Gln residues at site Asn 297. *Journal of Pharmaceutical Sciences*, *103*, 1613–1627.

An, Y., Zhang, Y., Mueller, H.-M., Shameem, M., & Chen, X. (2014). A new tool for monoclonal antibody analysis. *mAbs*, *6*, 879–893.

Anthony, R. M., Nimmerjahn, F., Ashline, D. J., Reinhold, V. N., Paulson, J. C., & Ravetch, J. V. (2008a). Recapitulation of IVIG anti-inflammatory activity with a recombinant IgG Fc. *Science (New York, N.Y.)*, *320*, 373–376.

Anthony, R. M., Wermeling, F., Karlsson, M. C., & Ravetch, J. V. (2008b). Identification of a receptor required for the anti-inflammatory activity of IVIG. *Proceedings of the National Academy of Sciences*, *105*, 19571–19578.

Anumula, K. R., & Dhume, S. T. (1998). High resolution and high sensitivity methods for oligosaccharide mapping and characterization by normal phase high performance liquid chromatography following derivatization with highly fluorescent anthranilic acid. *Glycobiology*, *8*, 685–694.

Aoyama, M., Hashii, N., Tsukimura, W., Osumi, K., Harazono, A., Tada, M., ... Ishii-Watabe, A. (2019). Effects of terminal galactose residues in mannose α1–6 arm of Fc-glycan on the effector functions of therapeutic monoclonal antibodies. *mAbs*, *11*, 826–836.

Bas, M., Terrier, A., Jacque, E., Dehenne, A., Pochet-Béghin, V., Beghin, C., ... Monnet, C. (2019). Fc sialylation prolongs serum half-life of therapeutic antibodies. *The Journal of Immunology*, ji1800896.

Borrok, M. J., Jung, S. T., Kang, T. H., Monzingo, A. F., & Georgiou, G. (2012). Revisiting the role of glycosylation in the structure of human IgG Fc. *ACS Chemical Biology*, 7, 1596–1602.

Boyd, P. N., Lines, A. C., & Patel, A. K. (1995). The effect of the removal of sialic acid, galactose and total carbohydrate on the functional activity of Campath-1H. *Molecular Immunology*, *32*, 1311–1318.

Brady, L. J., Valliere-Douglass, J., Martinez, T., & Balland, A. (2008). Molecular mass analysis of antibodies by on-line SEC-MS. *Journal of the American Society for Mass Spectrometry*, *19*, 502–509.

Carlsson, M., Braddock, M., Li, Y., Wang, J., Xu, W., White, N., ... Colice, G. (2019). Evaluation of antibody properties and clinically relevant immunogenicity, anaphylaxis, and hypersensitivity reactions in two phase III trials of tralokinumab in severe, uncontrolled asthma. *Drug Safety*, *42*, 769–784.

Chen, X., & Flynn, G. C. (2007). Analysis of N-glycans from recombinant immunoglobulin G by on-line reversed-phase high-performance liquid chromatography/mass spectrometry. *Analytical Biochemistry*, *370*, 147–161.

Chen, X., Liu, Y. D., & Flynn, G. C. (2008). The effect of Fc glycan forms on human IgG2 antibody clearance in humans. *Glycobiology*, *19*, 240–249.

Cho, I. H., Lee, N., Song, D., Jung, S. Y., Bou-Assaf, G., Sosic, Z., ... Lyubarskaya, Y. (2016). Evaluation of the structural, physicochemical, and biological characteristics of SB4, a biosimilar of etanercept. *mAbs*, *8*, 1136–1155.

Chung, C. H., Mirakhur, B., Chan, E., Le, Q.-T., Berlin, J., Morse, M., ... Platts-Mills, T. A. E. (2008). Cetuximab-induced anaphylaxis and IgE specific for galactose-alpha-1,3-galactose. *The New England journal of medicine*, *358*, 1109–1117.

Chung, S., Quarmby, V., Gao, X., Ying, Y., Lin, L., Reed, C., ... Song, A. (2012). Quantitative evaluation of fucose reducing effects in a humanized antibody on Fcγ receptor binding and antibody-dependent cell-mediated cytotoxicity activities. *mAbs*, *4*, 326–340.

Clarke, S. J., Smith, J. T., Gebbie, C., Sweeney, C., & Olszewski, N. (2009). A phase I, pharmacokinetic (PK), and preliminary efficacy assessment of ALD518, a humanized anti-IL-6 antibody, in patients with advanced cancer. *Journal of Clinical Oncology*, *27*, 3025.

D'atri, V., Fekete, S., Beck, A., Lauber, M., & Guillarme, D. (2017). Hydrophilic interaction chromatography hyphenated with mass spectrometry: a powerful analytical tool for the comparison of originator and biosimilar therapeutic monoclonal antibodies at the middle-up level of analysis. *Analytical Chemistry*, *89*, 2086–2092.

Daeron, M. (1997). Fc receptor biology. *Annual Review of Immunology*, *15*, 203–234.

Dashivets, T., Thomann, M., Rueger, P., Knaupp, A., Buchner, J., & Schlothauer, T. (2015). Multi-angle effector function analysis of human monoclonal IgG glycovariants. *PLoS One*, *10*, e0143520.

De Leoz, M. L. A., Duewer, D. L., Fung, A., Liu, L., Yau, H. K., Potter, O., ... Flaherty, R. (2020). *NIST interlaboratory study on glycosylation analysis of monoclonal antibodies: comparison of results from diverse analytical methods. Molecular Cellular Proteomics* (19, p. 11).

Dekkers, G., Treffers, L., Plomp, R., Bentlage, A. E. H., De Boer, M., Koeleman, C. A. M., ... Vidarsson, G. (2017). Decoding the human immunoglobulin G-glycan repertoire reveals a spectrum of Fc-receptor- and complement-mediated-effector activities. *Front Immunol*, *8*, 877.

Durocher, Y., & Butler, M. (2009). Expression systems for therapeutic glycoprotein production. *Current Opinion in Biotechnology*, *20*, 700–707.

Feige, M. J., Nath, S., Catharino, S. R., Weinfurtner, D., Steinbacher, S., & Buchner, J. (2009). Structure of the murine unglycosylated IgG1 Fc fragment. *Journal of Molecular Biology*, *391*, 599–608.

Ferrara, C., Brunker, P., Suter, T., Moser, S., Puntener, U., & Umana, P. (2006a). Modulation of therapeutic antibody effector functions by glycosylation engineering: influence of golgi enzyme localization domain and co-expression of heterologous beta1, 4-N-acetylglucosaminyltransferase III and golgi alpha-mannosidase II. *Biotechnology and Bioengineering*, *93*, 851–861.

Ferrara, C., Grau, S., Jager, C., Sondermann, P., Brunker, P., Waldhauer, I., ... Benz, J. (2011). *Unique carbohydrate-carbohydrate interactions are required for high affinity binding between FcgammaRIII and antibodies lacking core fucose, . Proceedings of the National Academy of Sciences of the United States of America* (108, pp. 12669–12674).

Ferrara, C., Stuart, F., Sondermann, P., Brunker, P., & Umana, P. (2006b). The carbohydrate at FcgammaRIIIa Asn-162. An element required for high affinity binding to non-fucosylated IgG glycoforms. *The Journal of Biological Chemistry*, *281*, 5032–5036.

Füssl, F., Trappe, A., Cook, K., Scheffler, K., Fitzgerald, O., & Bones, J. (2019). Comprehensive characterisation of the heterogeneity of adalimumab via charge variant analysis hyphenated on-line to native high resolution Orbitrap mass spectrometry. *mAbs*, *11*, 116–128.

Galili, U. (2013). Anti-Gal: an abundant human natural antibody of multiple pathogeneses and clinical benefits. *Immunology*, *140*, 1–11.

Gennaro, L. A., & Salas-Solano, O. (2008). On-line CE – LIF – MS technology for the direct characterization of N-linked glycans from therapeutic antibodies. *Analytical Chemistry*, *80*, 3838–3845.

Gerber, J. S., & Mosser, D. M. (2001). Stimulatory and inhibitory signals originating from the macrophage Fcgamma receptors. *Microbes and Infection/Institut Pasteur*, *3*, 131–139.

Ghaderi, D., Taylor, R. E., Padler-Karavani, V., Diaz, S., & Varki, A. (2010). Implications of the presence of N-glycolylneuraminic acid in recombinant therapeutic glycoproteins. *Nature Biotechnology*, *28*, 863–867.

Goetze, A. M., Liu, Y. D., Zhang, Z., Shah, B., Lee, E., Bondarenko, P. V., & Flynn, G. C. (2011). High-mannose glycans on the Fc region of therapeutic IgG antibodies increase serum clearance in humans. *Glycobiology*, *21*, 949–959.

Gstöttner, C., Nicolardi, S., Haberger, M., Reusch, D., Wuhrer, M., & Domínguez-Vega, E. (2020). Intact and subunit-specific analysis of bispecific antibodies by sheathless CE-MS. *Analytica Chimica Acta*, *1134*, 18–27.

Ha, S., Ou, Y., Vlasak, J., Li, Y., Wang, S., Vo, K., ... Zhang, N. (2011). Isolation and characterization of IgG1 with asymmetrical Fc glycosylation. *Glycobiology*, *21*, 1087–1096.

Haberger, M., Leiss, M., Heidenreich, A.-K., Pester, O., Hafenmair, G., Hook, M., ... Bulau, P. (2016). Rapid characterization of biotherapeutic proteins by size-exclusion chromatography coupled to native mass spectrometry. *mAbs*, *8*, 331–339.

Haselberg, R., De Vijlder, T., Heukers, R., Smit, M. J., Romijn, E. P., Somsen, G. W., & Dominguez-Vega, E. (2018). Heterogeneity assessment of antibody-derived therapeutics at the intact and middle-up level by low-flow sheathless capillary electrophoresis-mass spectrometry. *Analytica Chimica Acta*, *1044*, 181–190.

Hayes, J. M., Wormald, M. R., Rudd, P. M., & Davey, G. P. (2016). Fc gamma receptors: glycobiology and therapeutic prospects. *Journal of Inflammation Research*, *9*, 209–219.

Herter, S., Birk, M. C., Klein, C., Gerdes, C., Umana, P., & Bacac, M. (2014). Glycoengineering of therapeutic antibodies enhances monocyte/macrophage-mediated phagocytosis and cytotoxicity. *The Journal of Immunology*, *192*, 2252.

Higel, F., Seidl, A., Sörgel, F., & Friess, W. (2016). N-glycosylation heterogeneity and the influence on structure, function and pharmacokinetics of monoclonal antibodies and Fc fusion proteins. *European Journal of Pharmaceutics and Biopharmaceutics*, *100*, 94–100.

Hilger, C., Fischer, J., Wölbing, F., & Biedermann, T. (2019). Role and mechanism of galactose-alpha-1,3-galactose in the elicitation of delayed anaphylactic reactions to red meat. *Current Allergy and Asthma Reports*, *19*, 3.

Hilliard, M., Alley, W. R., JR., Mcmanus, C. A., Yu, Y. Q., Hallinan, S., Gebler, J., & Rudd, P. M. (2017). Glycan characterization of the NIST RM monoclonal antibody using a total analytical solution: from sample preparation to data analysis. *mAbs*, *9*, 1349–1359.

Hodoniczky, J., Zheng, Y. Z., & James, D. C. (2005). Control of recombinant monoclonal antibody effector functions by Fc N-glycan remodeling in vitro. *Biotechnology Progress*, *21*, 1644–1652.

Hokke, C. H., Bergwerff, A. A., Van Dedem, G. W., Van Oostrum, J., Kamerling, J. P., & Vliegenthart, J. F. (1990). Sialylated carbohydrate chains of recombinant human glycoproteins expressed in Chinese hamster ovary cells contain traces of N-glycolylneuraminic acid. *FEBS Letters*, *275*, 9–14.

Houde, D., Peng, Y., Berkowitz, S. A., & Engen, J. R. (2010). Post-translational modifications differentially affect IgG1 conformation and receptor binding. *Molecular Cellular Proteomics*, *9*, 1716.

Hristodorov, D., Fischer, R., Joerissen, H., Muller-Tiemann, B., Apeler, H., & Linden, L. (2013). Generation and comparative characterization of glycosylated and aglycosylated human IgG1 antibodies. *Molecular Biotechnology*, *53*, 326–335.

Huang, L., Biolsi, S., Bales, K. R., & Kuchibhotla, U. (2006). Impact of variable domain glycosylation on antibody clearance: an LC/MS characterization. *Analytical Biochemistry*, *349*, 197–207.

Huber, R., Deisenhofer, J., Colman, P. M., Matsushima, M., & Palm, W. (1976). Crystallographic structure studies of an IgG molecule and an Fc fragment. *Nature*, *264*, 415–420.

Hülsmeier, A. J., Welti, M., & Hennet, T. (2011). Glycoprotein maturation and the UPR. *Methods in Enzymology*, *491*, 163–182.

Iida, S., Kuni-Kamochi, R., Mori, K., Misaka, H., Inoue, M., Okazaki, A., ... Satoh, M. (2009). Two mechanisms of the enhanced antibody-dependent cellular cytotoxicity (ADCC) efficacy of non-fucosylated therapeutic antibodies in human blood. *BMC Cancer*, *9*, 58.

Jefferis, R. (2005). Glycosylation of recombinant antibody therapeutics. *Biotechnology Progress*, *21*, 11–16.

Jefferis, R. (2006). A sugar switch for anti-inflammatory antibodies. *Nature Biotechnology*, *24*, 1230–1231.

Jefferis, R. (2007). Antibody therapeutics: isotype and glycoform selection. *Expert Opinion on Biological Therapy*, 7, 1401–1413.

Jefferis, R. (2009). Recombinant antibody therapeutics: the impact of glycosylation on mechanisms of action. *Trends in Pharmacological Sciences*, *30*, 356–362.

Jefferis, R. (2012). Isotype and glycoform selection for antibody therapeutics. *Archives of Biochemistry and Biophysics*, *526*, 159–166.

Jefferis, R. (2016). Posttranslational modifications and the immunogenicity of biotherapeutics. *Journal of Immunology Research*, *2016*, 5358272.

Jenkins, N., Parekh, R. B., & James, D. C. (1996). Getting the glycosylation right: implications for the biotechnology industry. *Nature Biotechnology*, *14*, 975–981.

Jiang, X. R., Song, A., Bergelson, S., Arroll, T., Parekh, B., May, K., ... Schenerman, M. (2011). Advances in the assessment and control of the effector functions of therapeutic antibodies. *Nature Reviews Drug Discovery*, *10*, 101–111.

Jung, S. T., Kang, T. H., Kelton, W., & Georgiou, G. (2011). Bypassing glycosylation: engineering aglycosylated full-length IgG antibodies for human therapy. *Current Opinion in Biotechnology*, *22*, 858–867.

Jung, S. T., Reddy, S. T., Kang, T. H., Borrok, M. J., Sandlie, I., Tucker, P. W., & Georgiou, G. (2010). Aglycosylated IgG variants expressed in bacteria that selectively bind FcγRI potentiate tumor cell killing by monocyte-dendritic cells. *Proceedings of the National Academy of Sciences*, *107*, 604–609.

Kamoda, S., Ishikawa, R., & Kakehi, K. (2006). Capillary electrophoresis with laser-induced fluorescence detection for detailed studies on N-linked oligosaccharide profile of therapeutic recombinant monoclonal antibodies. *Journal of Chromatography. A*, *1133*, 332–339.

Kamoda, S., Nomura, C., Kinoshita, M., Nishiura, S., Ishikawa, R., Kakehi, K., ... Hayakawa, T. (2004). Profiling analysis of oligosaccharides in antibody pharmaceuticals by capillary electrophoresis. *Journal of Chromatography. A*, *1050*, 211–216.

Kanda, Y., Yamada, T., Mori, K., Okazaki, A., Inoue, M., Kitajima-Miyama, K., ... Satoh, M. (2006). Comparison of biological activity among nonfucosylated therapeutic

IgG1 antibodies with three different N-linked Fc oligosaccharides: the high-mannose, hybrid, and complex types. *Glycobiology*, *17*, 104–118.

Kaneko, Y., Nimmerjahn, F., & Ravetch, J. V. (2006). Anti-inflammatory activity of immunoglobulin G resulting from Fc sialylation. *Science (New York, N.Y.)*, *313*, 670–673.

Kaur, H. (2021). Characterization of glycosylation in monoclonal antibodies and its importance in therapeutic antibody development. *Critical Reviews in Biotechnology*, 1–23.

Kaur, H. (2021). Stability testing in monoclonal antibodies. *Critical Reviews in Biotechnology*. Available from https://doi.org/10.1080/07388551.2021.1874281.

Kayser, V., Chennamsetty, N., Voynov, V., Forrer, K., Helk, B., & Trout, B. L. (2011). Glycosylation influences on the aggregation propensity of therapeutic monoclonal antibodies. *Biotechnology Journal*, *6*, 38–44.

Keser, T., Pavić, T., Lauc, G., & Gornik, O. (2018). Comparison of 2-aminobenzamide, procainamide and RapiFluor-MS as derivatizing agents for high-throughput HILIC-UPLC-FLR-MS N-glycan analysis. *Frontiers in chemistry*, *6*, 324.

Khatri, K., Klein, J. A., Haserick, J. R., Leon, D. R., Costello, C. E., Mccomb, M. E., & Zaia, J. (2017). Microfluidic capillary electrophoresis-mass spectrometry for analysis of monosaccharides, oligosaccharides, and glycopeptides. *Analytical Chemistry*, *89*, 6645–6655.

Kim, J.-K., Tsen, M.-F., Ghetie, V., & Sally Ward, E. (1994). Identifying amino acid residues that influence plasma clearance of murine IgG1 fragments by site-directed mutagenesis. *European Journal of Immunology*, *24*, 542–548.

Krapp, S., Mimura, Y., Jefferis, R., Huber, R., & Sondermann, P. (2003). Structural analysis of human IgG-Fc glycoforms reveals a correlation between glycosylation and structural integrity. *Journal of Molecular Biology*, *325*, 979–989.

Lammerts Van Bueren, J. J., Rispens, T., Verploegen, S., Van Der Palen-Merkus, T., Stapel, S., Workman, L. J., … Parren, P. W. (2011). Anti-galactose-alpha-1,3-galactose IgE from allergic patients does not bind alpha-galactosylated glycans on intact therapeutic antibody Fc domains. *Nature Biotechnology*, *29*, 574–576.

Larsen, R. D., Rajan, V. P., Ruff, M. M., Kukowska-Latallo, J., Cummings, R. D., & Lowe, J. B. (1989). Isolation of a cDNA encoding a murine UDPgalactose:beta-D-galactosyl-1,4-N-acetyl-D-glucosaminide alpha-1,3-galactosyltransferase: expression cloning by gene transfer. *Proceedings of the National Academy of Sciences of the United States of America*, *86*, 8227–8231.

Li, T., Dilillo, D. J., Bournazos, S., Giddens, J. P., Ravetch, J. V., & Wang, L. X. (2017). Modulating IgG effector function by Fc glycan engineering. *Proceedings of the National Academy of Sciences of the United States of America*, *114*, 3485–3490.

Lim, A., Reed-Bogan, A., & Harmon, B. J. (2008). Glycosylation profiling of a therapeutic recombinant monoclonal antibody with two N-linked glycosylation sites using liquid chromatography coupled to a hybrid quadrupole time-of-flight mass spectrometer. *Analytical Biochemistry*, *375*, 163–172.

Lin, C. W., Tsai, M. H., Li, S. T., Tsai, T. I., Chu, K. C., Liu, Y. C., … Wong, C. H. (2015). A common glycan structure on immunoglobulin G for enhancement of effector functions. *Proceedings of the National Academy of Sciences of the United States of America*, *112*, 10611–10616.

Lippold, S., Nicolardi, S., Domínguez-Vega, E., Heidenreich, A.-K., Vidarsson, G., Reusch, D., … Falck, D. (2019a). Glycoform-resolved FcγRIIIa affinity chromatography-mass spectrometry. *mAbs*, *11*, 1191–1196.

Lippold, S., Nicolardi, S., Wuhrer, M., & Falck, D. (2019b). Proteoform-resolved FcγRIIIa binding assay for Fab glycosylated monoclonal antibodies achieved by affinity chromatography mass spectrometry of Fc moieties. *Frontiers in chemistry*, 7, 698.

Liu, L. (2018). Pharmacokinetics of monoclonal antibodies and Fc-fusion proteins. *Protein & cell*, *9*, 15–32.

Liu, L., Stadheim, A., Hamuro, L., Pittman, T., Wang, W., Zha, D., ... Prueksaritanont, T. (2011). Pharmacokinetics of IgG1 monoclonal antibodies produced in humanized Pichia pastoris with specific glycoforms: a comparative study with CHO produced materials. *Biologicals: Journal of the International Association of Biological Standardization*, *39*, 205–210.

Ludger Product Guide for LudgerLiberate™ Orela Glycan Release Kit. https://www.ludger.com/docs/products/ll/ll-orela-a2/ludger-ll-orela-a2-guide.pdf.

Lund, J., Takahashi, N., Pound, J. D., Goodall, M., & Jefferis, R. (1996). Multiple interactions of IgG with its core oligosaccharide can modulate recognition by complement and human Fc gamma receptor I and influence the synthesis of its oligosaccharide chains. *The Journal of Immunology*, *157*, 4963–4969.

Lund, J., Takahashi, N., Pound, J. D., Goodall, M., Nakagawa, H., & Jefferis, R. (1995). Oligosaccharide-protein interactions in IgG can modulate recognition by Fc gamma receptors. *The FASEB Journal*, *9*, 115–119.

Martin, A. C. (1996). Accessing the Kabat antibody sequence database by computer. *Proteins*, *25*, 130–133.

Melmer, M., Stangler, T., Schiefermeier, M., Brunner, W., Toll, H., Rupprechter, A., ... Premstaller, A. (2010). HILIC analysis of fluorescence-labeled N-glycans from recombinant biopharmaceuticals. *Analytical and Bioanalytical Chemistry*, *398*, 905–914.

Mimura, Y., Church, S., Ghirlando, R., Ashton, P. R., Dong, S., Goodall, M., ... Jefferis, R. (2000). The influence of glycosylation on the thermal stability and effector function expression of human IgG1-Fc: properties of a series of truncated glycoforms. *Molecular Immunology*, *37*, 697–706.

Moroncini, G., Maccaroni, E., Fiordoliva, I., Pellei, C., Gabrielli, A., & Berardi, R. (2018). Developments in the management of advanced soft-tissue sarcoma—olaratumab in context. *OncoTargets and Therapy*, *11*, 833–842.

Naso, M. F., Tam, S. H., Scallon, B. J., & Raju, T. S. (2010). Engineering host cell lines to reduce terminal sialylation of secreted antibodies. *mAbs*, *2*, 519–527.

Nimmerjahn, F., Anthony, R. M., & Ravetch, J. V. (2007). Agalactosylated IgG antibodies depend on cellular Fc receptors for in vivo activity. *Proceedings of the National Academy of Sciences of the United States of America*, *104*, 8433–8437.

Nimmerjahn, F., & Ravetch, J. V. (2008a). Anti-inflammatory actions of intravenous immunoglobulin. *Annual Review of Immunology*, *26*, 513–533.

Nimmerjahn, F., & Ravetch, J. V. (2008b). Fcgamma receptors as regulators of immune responses. *Nature Reviews. Immunology*, *8*, 34–47.

Noguchi, A., Mukuria, C. J., Suzuki, E., & Naiki, M. (1996). Failure of human immunoresponse to N-glycolylneuraminic acid epitope contained in recombinant human erythropoietin. *Nephron*, *72*, 599–603.

Nose, M., & Wigzell, H. (1983). Biological significance of carbohydrate chains on monoclonal antibodies. *Proceedings of the National Academy of Sciences*, *80*, 6632.

Okazaki, A., Shoji-Hosaka, E., Nakamura, K., Wakitani, M., Uchida, K., Kakita, S., ... Shitara, K. (2004). Fucose depletion from human IgG1 oligosaccharide enhances binding enthalpy and association rate between IgG1 and FcgammaRIIIa. *Journal of Molecular Biology*, *336*, 1239–1249.

Padler-Karavani, V., Yu, H., Cao, H., Chokhawala, H., Karp, F., Varki, N., ... Varki, A. (2008). Diversity in specificity, abundance, and composition of anti-Neu5Gc antibodies in normal humans: potential implications for disease. *Glycobiology*, *18*, 818–830.

Peipp, M., Lammerts Van Bueren, J. J., Schneider-Merck, T., Bleeker, W. W., Dechant, M., Beyer, T., ... Valerius, T. (2008). Antibody fucosylation differentially impacts cytotoxicity mediated by NK and PMN effector cells. *Blood*, *112*, 2390–2399.

Pereira, N. A., Chan, K. F., Lin, P. C., & Song, Z. (2018). The "less-is-more" in therapeutic antibodies: afucosylated anti-cancer antibodies with enhanced antibody-dependent cellular cytotoxicity. *MAbs*, *10*, 693–711.

Prozyme Pro-Link Extender™ Kit for Complex O-Linked Glycans. https://prozymetoo.myshopify.com/collections/o-glycans-kits/products/gk80115.

Qian, J., Liu, T., Yang, L., Daus, A., Crowley, R., & Zhou, Q. (2007). Structural characterization of N-linked oligosaccharides on monoclonal antibody cetuximab by the combination of orthogonal matrix-assisted laser desorption/ionization hybrid quadrupole-quadrupole time-of-flight tandem mass spectrometry and sequential enzymatic digestion. *Analytical Biochemistry*, *364*, 8–18.

Radaev, S., & Sun, P. (2002). Recognition of immunoglobulins by Fcgamma receptors. *Molecular Immunology*, *38*, 1073–1083.

Raju, S. (2003). Glycosylation variations with expression systems. *BioProcess International*, *1*, 44–53.

Raju, T. S., Briggs, J. B., Borge, S. M., & Jones, A. J. S. (2000). Species-specific variation in glycosylation of IgG: evidence for the species-specific sialylation and branch-specific galactosylation and importance for engineering recombinant glycoprotein therapeutics. *Glycobiology*, *10*, 477–486.

Ravetch, J. V. (1997). Fc receptors. *Current Opinion in Immunology*, *9*, 121–125.

Ravetch, J. V., & Bolland, S. (2001). IgG Fc receptors. *Annual Review of Immunology*, *19*, 275–290.

Reiding, K. R., Blank, D., Kuijper, D. M., Deelder, A. M., & Wuhrer, M. (2014). High-throughput profiling of protein N-glycosylation by MALDI-TOF-MS employing linkage-specific sialic acid esterification. *Analytical Chemistry*, *86*, 5784–5793.

Reusch, D., Haberger, M., Falck, D., Peter, B., Maier, B., Gassner, J., ... Wuhrer, M. (2015a). Comparison of methods for the analysis of therapeutic immunoglobulin G Fc-glycosylation profiles—part 2: mass spectrometric methods. *MAbs*, *7*, 732–742.

Reusch, D., Haberger, M., Kailich, T., Heidenreich, A.-K., Kampe, M., Bulau, P., & Wuhrer, M. (2014). High-throughput glycosylation analysis of therapeutic immunoglobulin G by capillary gel electrophoresis using a DNA analyzer. *mAbs*, *6*, 185–196.

Reusch, D., Haberger, M., Maier, B., Maier, M., Kloseck, R., Zimmermann, B., ... Wuhrer, M. (2015b). Comparison of methods for the analysis of therapeutic immunoglobulin G Fc-glycosylation profiles—part 1: separation-based methods. *MAbs*, *7*, 167–179.

Reusch, D., & Tejada, M. L. (2015). Fc glycans of therapeutic antibodies as critical quality attributes. *Glycobiology*, *25*, 1325–1334.

Ritamo, I., Cloutier, M., Valmu, L., Neron, S., & Rabina, J. (2014). Comparison of the glycosylation of in vitro generated polyclonal human IgG and therapeutic immunoglobulins. *Molecular Immunology*, *57*, 255–262.

Rogers, R. S., Nightlinger, N. S., Livingston, B., Campbell, P., Bailey, R., & Balland, A. (2015). Development of a quantitative mass spectrometry multi-attribute method for characterization, quality control testing and disposition of biologics. *MAbs*, *7*, 881–890.

Rogstad, S., Yan, H., Wang, X., Powers, D., Brorson, K., Damdinsuren, B., & Lee, S. (2019). Multi-attribute method for quality control of therapeutic proteins. *Analytical Chemistry*, *91*, 14170–14177.

Ruhaak, L. R., Zauner, G., Huhn, C., Bruggink, C., Deelder, A. M., & Wuhrer, M. (2010). Glycan labeling strategies and their use in identification and quantification. *Analytical and Bioanalytical Chemistry*, *397*, 3457–3481.

Scallon, B. J., Tam, S. H., Mccarthy, S. G., Cai, A. N., & Raju, T. S. (2007). Higher levels of sialylated Fc glycans in immunoglobulin G molecules can adversely impact functionality. *Mol Immunol, 44*, 1524–1534.

Schiestl, M., Stangler, T., Torella, C., Čepeljnik, T., Toll, H., & Grau, R. (2011). Acceptable changes in quality attributes of glycosylated biopharmaceuticals. *Nature Biotechnology, 29*, 310–312.

Shatz, W., Chung, S., Li, B., Marshall, B., Tejada, M., Phung, W., ... Scheer, J. M. (2013). Knobs-into-holes antibody production in mammalian cell lines reveals that asymmetric afucosylation is sufficient for full antibody-dependent cellular cytotoxicity. *mAbs, 5*, 872–881.

Sheeley, D. M., Merrill, B. M., & Taylor, L. C. (1997). Characterization of monoclonal antibody glycosylation: comparison of expression systems and identification of terminal alpha-linked galactose. *Analytical Biochemistry, 247*, 102–110.

Shields, R. L., Lai, J., Keck, R., O'connell, L. Y., Hong, K., Meng, Y. G., ... Presta, L. G. (2002). Lack of fucose on human IgG1 N-linked oligosaccharide improves binding to human Fcgamma RIII and antibody-dependent cellular toxicity. *The Journal of Biological Chemistry, 277*, 26733–26740.

Shinkawa, T., Nakamura, K., Yamane, N., Shoji-Hosaka, E., Kanda, Y., Sakurada, M., ... Shitara, K. (2003). The absence of fucose but not the presence of galactose or bisecting N-acetylglucosamine of human IgG1 complex-type oligosaccharides shows the critical role of enhancing antibody-dependent cellular cytotoxicity. *The Journal of Biological Chemistry, 278*, 3466–3473.

Sondermann, P., Kaiser, J., & Jacob, U. (2001). Molecular basis for immune complex recognition: a comparison of Fc-receptor structures. *Journal of Molecular Biology, 309*, 737–749.

Sörman, A., Zhang, L., Ding, Z., & Heyman, B. (2014). How antibodies use complement to regulate antibody responses. *Molecular Immunology, 61*, 79–88.

Spiess, M. (1990). The asialoglycoprotein receptor: a model for endocytic transport receptors. *Biochemistry, 29*, 10009–10018.

Tangvoranuntakul, P., Gagneux, P., Diaz, S., Bardor, M., Varki, N., Varki, A., & Muchmore, E. (2003). Human uptake and incorporation of an immunogenic nonhuman dietary sialic acid. *Proc Natl Acad Sci U S A, 100*, 12045–12050.

Tao, M. H., & Morrison, S. L. (1989). Studies of aglycosylated chimeric mouse-human IgG. Role of carbohydrate in the structure and effector functions mediated by the human IgG constant region. *Journal of Immunology, 143*, 2595–2601.

Taylor, R. E., Gregg, C. J., Padler-Karavani, V., Ghaderi, D., Yu, H., Huang, S., ... Varki, A. (2010). Novel mechanism for the generation of human xeno-autoantibodies against the nonhuman sialic acid N-glycolylneuraminic acid. *Journal of Experimental Medicine, 207*, 1637–1646.

Temming, A. R., Dekkers, G., Van De Bovenkamp, F. S., Plomp, H. R., Bentlage, A. E. H., Szittner, Z., ... Vidarsson, G. (2019). Human DC-SIGN and CD23 do not interact with human IgG. *Scientific Reports, 9*, 9995.

Thomann, M., Malik, S., Kuhne, F., Avenal, C., Plath, F., Bonnington, L., ... Cymer, F. (2019). Effects of sialic acid linkage on antibody-fragment crystallizable receptor binding and antibody dependent cytotoxicity depend on levels of fucosylation/bisecting. *Bioanalysis, 11*, 1437–1449.

Thomann, M., Reckermann, K., Reusch, D., Prasser, J., & Tejada, M. L. (2016). Fc-galactosylation modulates antibody-dependent cellular cytotoxicity of therapeutic antibodies. *Molecular Immunology, 73*, 69–75.

Thomann, M., Schlothauer, T., Dashivets, T., Malik, S., Avenal, C., Bulau, P., ... Reusch, D. (2015). In vitro glycoengineering of IgG1 and its effect on Fc receptor binding and ADCC activity. *PLoS One, 10*, e0134949.

Tsuchiya, N., Endo, T., Matsuta, K., Yoshinoya, S., Aikawa, T., Kosuge, E., ... Kobata, A. (1989). Effects of galactose depletion from oligosaccharide chains on immunological activities of human IgG. *The Journal of Rheumatology*, *16*, 285–290.

Turyan, I., Hronowski, X., Sosic, Z., & Lyubarskaya, Y. (2014). Comparison of two approaches for quantitative O-linked glycan analysis used in characterization of recombinant proteins. *Analytical Biochemistry*, *446*, 28–36.

Umana, P., Jean-Mairet, J., Moudry, R., Amstutz, H., & Bailey, J. E. (1999). Engineered glycoforms of an antineuroblastoma IgG1 with optimized antibody-dependent cellular cytotoxic activity. *Nature Biotechnology*, *17*, 176–180.

Van Den Steen, P., Rudd, P. M., Dwek, R. A., & Opdenakker, G. (1998). Concepts and principles of O-linked glycosylation. *Critical Reviews in Biochemistry and Molecular Biology*, *33*, 151–208.

Van Tine, B. A., Govindarajan, R., Attia, S., Somaiah, N., Barker, S. S., Shahir, A., ... Tap, W. D. (2019). Incidence and management of olaratumab infusion-related reactions. *Journal of Oncology Practice*, *15*, e925–e933.

Varki, A. (2007). Glycan-based interactions involving vertebrate sialic-acid-recognizing proteins. *Nature*, *446*, 1023–1029.

Wada, R., Matsui, M., & Kawasaki, N. (2019). Influence of N-glycosylation on effector functions and thermal stability of glycoengineered IgG1 monoclonal antibody with homogeneous glycoforms. *mAbs*, *11*, 350–372.

Wagner-Rousset, E., Bednarczyk, A., Bussat, M.-C., Colas, O., Corvaïa, N., Schaeffer, C., ... Beck, A. (2008). The way forward, enhanced characterization of therapeutic antibody glycosylation: comparison of three level mass spectrometry-based strategies. *Journal of Chromatography B*, *872*, 23–37.

Wawrzynczak, E. J., Cumber, A. J., Parnell, G. D., Jones, P. T., & Winter, G. (1992). Blood clearance in the rat of a recombinant mouse monoclonal antibody lacking the N-linked oligosaccharide side chains of the CH2 domains. *Molecular Immunology*, *29*, 213–220.

Wawrzynczak, E. J., Parnell, G. D., Cumber, A. J., Jones, P. T., & Winter, G. (1989). Blood clearance in the mouse of an aglycosyl recombinant monoclonal antibody. *Biochemical Society Transactions*, *17*, 1061–1062.

Wright, A., & Morrison, S. L. (1994). Effect of altered CH2-associated carbohydrate structure on the functional properties and in vivo fate of chimeric mouse-human immunoglobulin. *Journal of Experimental Medicine*, *180*, 1087–1096.

Wright, A., & Morrison, S. L. (1998). Effect of C2-associated carbohydrate structure on Ig effector function: studies with chimeric mouse-human IgG1 antibodies in glycosylation mutants of chinese hamster ovary cells. *The Journal of Immunology*, *160*, 3393–3402.

Wright, A., Sato, Y., Okada, T., Chang, K. H., Endo, T., & Morrison, S. L. (2000). In vivo trafficking and catabolism of IgG1 antibodies with Fc associated carbohydrates of differing structure. *Glycobiology*, *10*, 1347–1355.

Xie, H., Chakraborty, A., Ahn, J., Yu, Y. Q., Dakshinamoorthy, D. P., Gilar, M., ... Mazzeo, J. R. (2010). Rapid comparison of a candidate biosimilar to an innovator monoclonal antibody with advanced liquid chromatography and mass spectrometry technologies. *mAbs*, *2*, 379–394.

Yu, C., Gao, K., Zhu, L., Wang, W., Wang, L., Zhang, F., ... Wang, J. (2016). At least two Fc Neu5Gc residues of monoclonal antibodies are required for binding to anti-Neu5Gc antibody. *Scientific Reports*, *6*, 20029.

Yu, M., Brown, D., Reed, C., Chung, S., Lutman, J., Stefanich, E., ... Bayer, R. (2012). Production, characterization and pharmacokinetic properties of antibodies with N-linked mannose-5 glycans. *mAbs*, *4*, 475–487.

Zeck, A., Pohlentz, G., Schlothauer, T., Peter-Katalinić, J., & Regula, J. T. (2011). Cell type-specific and site directed N-glycosylation pattern of FcγRIIIa. *Journal of Proteome Research, 10*, 3031–3039.

Zhang, P., Woen, S., Wang, T., Liau, B., Zhao, S., Chen, C., ... Rudd, P. M. (2016). Challenges of glycosylation analysis and control: an integrated approach to producing optimal and consistent therapeutic drugs. *Drug Discovery Today, 21*, 740–765.

Zheng, K., Bantog, C., & Bayer, R. (2011). The impact of glycosylation on monoclonal antibody conformation and stability. *mAbs, 3*, 568–576.

Zhu, A., & Hurst, R. (2002). Anti-N-glycolylneuraminic acid antibodies identified in healthy human serum. *Xenotransplantation, 9*, 376–381.

CHAPTER 4

Primary structure and peptide mapping

Thomas Flad[1] and Chris Chumsae[2]
[1]Protagen Protein Services GmbH, Heilbronn, Germany
[2]Bristol Myers Squibb, Process Development Analytics, Devens, MA, United States

4.1 Introduction

The primary structure of a monoclonal antibody (mAb), that is, its amino acid sequence lays the basis for the molecule properties and therefore is of large importance. For biotechnology products, the evaluation and control of primary sequence is described by the *International Council for Harmonization of Technical Requirements for Pharmaceuticals for Human Use* (ICH) within guideline Q6B (Specifications: Test Procedures and Acceptance Criteria for Biotechnological/Biological Products).

Conventional IgG molecules contain two identical heavy and light chains with ~450 and ~210 amino acids, respectively. Newer bi-specific mAbs typically contain two different heavy and light chains, which makes the control of the correct amino acid sequence more complex.

For analytical assessment of correct primary structure traditionally Edman sequencing has been applied, where amino acids are sequentially determined after chemical cleavage of the N-terminus (Fruton, 1992). Although this technology has some advantages such as clear discrimination between the isobaric amino acids leucine and isoleucine, there are several limitations. These are the limited number of amino acids, which can be determined (typically only up to 30 aa), the necessity for a free N-terminus of the target protein to start the degradation process as well as limitations to detect modifications of amino acid side chains. Therefore, mass spectrometry (MS) has become the method of choice for simultaneous analysis of amino acid sequence and posttranslational modifications. MS is capable to analyze proteins in the intact form (Tassi et al., 2018) as well as after proteolytic digestion and has become an essential tool in the biopharmaceutical industry for the development and quality control and will now be described in more detail.

Monoclonal Antibodies
DOI: https://doi.org/10.1016/B978-0-12-822318-5.00004-1
Copyright © 2021 Elsevier Inc.
All rights reserved.

4.2 Peptide mapping

Peptide mapping (PM) is widely applied in the biopharmaceutical industry since many years (Dong, 1992; Dong & Tran, 1990). Various protein-based drugs have been analyzed by this technique such as somatotropin (Dougherty, Snyder, Sinclair, & Robins, 1990), granulocyte colony-stimulating factor (Jones, Merewether, Clogston, & Lu, 1994), or mAbs (Kannan et al., 1997). A PM approach typically consists of following steps, which will subsequently be described in more detail:

- Sample preparation of the protein
- Cleavage of the protein into peptides by endoproteases
- Instrumental analysis of the resulting peptide pool for identity testing, extended characterization and comparability, or multiattribute method (MAM) approaches

4.2.1 Sample preparation

Sample preparation before endoproteolytical cleavage depends on the analytical question, which has to be addressed and basically includes following steps:

- Adjustment of appropriate pH value of the sample
- Removal of components disturbing subsequent instrumental analysis
- Optional de-glycosylation
- Protein denaturation/reduction and alkylation of cysteine-based disulfide bridges

Adjustment of suitable pH value depends on solubility range of the mAb itself, pH optimum of the later used endoprotease as well as avoidance of artificially induced side chain variations such as pH-induced asparagine deamidation to aspartic and isoaspartic acid (Kori, Patel, Neill, & Liu, 2015; Pace, Wong, Zhang, Kao, & Wang, 2013; Robinson, Scotchler, & McKerrow, 1973). Mostly, the selection of pH value is therefore a compromise addressing these different aspects. For mAb products, for example, a typical pH range for subsequent tryptic digestion is at about pH 6.0–8.0.

4.2.1.1 Disturbing components

mAb products are mostly present in a formulation buffer-containing different salts, sometimes amino acids (such as methionine), sometimes sugar alcohols (such as mannitol), and in many cases detergents as solubilization agents. As the mAb products are subsequently injected into patients, only a limited number of detergents for solubilization are allowed. Commonly, either polysorbate 20 (PS20) or polysorbate 80 (PS80) are used. Among the typical formulation components mostly the detergents can cause instrumental issues in particular for MS. Therefore, the amount of detergent has to be diminished either by dilution of the sample (if the protein concentration is high enough) or by sample rebuffering.

4.2.1.2 De-glycosylation

Most of the current mAb products have only one N-glycosylation site in the Fc part of the heavy chain. For analysis of glycosylation heterogeneity on the glycopeptide level (see Chapter 3: Glycosylation Analysis), no de-glycosylation is performed. For some analytical questions (e.g., for sequence variant analysis) the N-glycans are removed by enzymatic treatment with peptide N-glycosidase F (PNGase F) in order to reduce the heterogeneity of the peptide pool (Medzihradszky, 2008).

4.2.1.3 Protein denaturation/reduction

Protein denaturation and subsequent reduction and alkylation of cysteines are performed in most cases, as the subsequent proteolytic cleavage is more efficient and furthermore through blocking of free thiols by alkylation no unwanted di- or oligo-merization of cysteine containing peptides occurs. No reduction or alkylation of cysteines is performed for evaluation of correct disulfide bridging and of possible cysteine-related mAb modifications such as unwanted free thiols, trisulfides, cysteinylation, or thioether structures (Baker et al., 2019; Cheng et al., 2017; Kita, Ponniah, Nowak, & Liu, 2016).

Denaturation is performed by elevated temperature as well as by usage of denaturing agents, mostly guanidine HCl or urea. Again a compromise between mild conditions for avoidance of chemically induced artifacts and efficient denaturation for maximum digestion is applied (Liu, Gaza-Bulseco, & Sun, 2006).

4.2.2 Endoproteases

For subsequent instrumental analysis the prepared protein is cleaved by particular endoproteases. The aim is to ideally generate a peptide pool with peptides in the size range of minimum four amino acids (to facilitatate specificity) and maximum 30–40 amino acids (to still maintain manageability by MS). To realize this, a toolbox of particular endoproteases is applied.

The following table lists the most commonly used endoproteases for PM and shows respective cleavage specificity (also refer to Swiss Institute of Bioinformatics Resource Portal Expasy) (Table 4.1).

Cleavage occurs between P1 and P1′ position. For most proteases not only P1 and P1′ position have influences on cleavage but also other amino acid positions near the cleavage site. For example, Trypsin as one of the

Table 4.1 List of important endoproteases used for peptide mapping with cleavage specificity.

Enzyme	P3	P2	P1	P1′	P2′
Trypsin basic rule	–	–	**K** or **R**	not P	–
Exceptions 1–2 (cleavage occurs)	–	W	**K**	P	–
	–	M	**R**	P	–
Exceptions 3–6 (no cleavage)	–	C or D	K	D	–
	–	C	K	H or Y	–
	–	C	R	K	–
	–	R	R	H or R	–
LysC	–	–	**K**	–	–
Asp-N	–	–	–	**D**	–
Arg-C	–	–	–	**R**	–
Glu-C (pH4)	–	–	–	**E**	–
Glu-C (pH8)	–	–	–	**E** or **D**	–
Chymotrypsin (main specificity)	–	–	**F** or **Y**	Not P	–
	–	–	**W**	Not M or P	–
Chymotrypsin (lower specificity)	–	–	**L**	Not P	–
	–	–	**M**	Not P or Y	–
	–	–	**H**	Not D, M, P, or W	–
Pepsin (pH1.3)	Not H, K or R	Not P	Not R	**F** or **L**	Not P
	Not H, K, or R	Not P	**F** or **L**	–	Not P
Pepsin (pH > 2)	Not H, K, or R	Not P	Not R	**F**, **L**, **W**, or **Y**	Not P
	Not H, K, or R	Not P	**F**, **L**, **W**, or **Y**	–	Not P
Chemical cleavage[a]					
CNBr	–	–	**M**	–	–
BNPS-Skatole	–	–	**W**	–	–

Following scheme for cleavage nomenclature is applied (Schechter & Berger, 1968).
N-Term ----- P3-P2-P1 -$\downarrow$- P1′-P2′-P3′ -----C-Term.
$\downarrow$ = cleavage site.
[a]Crimmins, Mische, and Denslow (2005).
Bold values indicates the cleavage specificity.

most prominent proteases used for PM typically cleaves after lysine (K) or arginine (R) if no proline (P) is in position P1′. However, as an exception, Trypsin can cleave after lysine (K) even with proline in position P1′ if tryptophane (W) is present at position P2 and also after arginine (R) if methionine (M) is present at position P2. Moreover, even without proline (P) in position P1′ Trypsin does not cleave with some combinations of amino acids in P2 and P1′ (listed as exceptions 3–6). In general the cleavage rules indicate the majority of cleavage events. However, different cleavages not following the rules (so-called mis-cleavages) frequently occur in PM approaches, and the amount of these mis-cleavages is dependent on the chemical conditions of the reaction (pH, temperature, buffer, etc.).

The respective proteases can be either used for single-digestion approaches or in combination as multiple proteases approach (see Section 4.3.5.2) in relation to the desired peptide pool for subsequent analysis. Besides cleaving specificity of the used endoproteases it is of importance, that different enzyme production lots can be supplied on a constant quality as these are critical reagents for PM and essentially influence quality and reproducibility of the experiments.

The digestion conditions including pH value, enzyme to substrate ratio, temperature, and time are again a compromise for facilitating maximum digestion with minimum introduction of artificial modifications. A systematic approach for optimization of the sample preparation and digestion conditions for a mAb product has been described by Mouchahoir and Schiel (2018).

Typically the resulting peptide pools are then separated by reversed phase liquid chromatography (RP-HPLC or RP-UPLC) and the eluting peptides analyzed by UV absorption and depending on the analytical scope also by MS, which will be illustrated in more detail below.

The resulting PM analytical data set is used within the biopharmaceutical industry for following different purposes.

4.2.3 Identity testing

The proteolytically generated peptide pool is unique for every protein based on the specific amino acid sequence and therefore is predestinated to be used for identity testing (ICH Q6B 4.1.2. Identity). Identity testing is applied to avoid unwanted mix-up of different protein drugs produced within one production site. The more comparable mAb products are produced at one site the more specific the method for identity testing has to be. Specificity using MS as the final detector for readout of the peptide pool is highest as every peptide is unambiguously identified by exact mass of the intact peptide as well as of the peptide fragments generated in the mass spectrometer. However, a mass spectrometer is a very expensive detector and the MS data generated are very complex. Therefore, in most cases the LC-UV-MS setup for PM is only used for the development of an identity test for assignment of UV-detected elution peaks and optimization of sample preparation and LC settings. For the later method validation and routine release identity testing only the LC-UV setup is used without MS detection.

4.2.4 Characterization and comparability

Analysis of the PM pool is very often basis for characterization (ICH Q6B) and comparability analysis (ICH Q5E) of protein-based drug candidates.

Assessment of correct primary structure is an essential part of ICH Q6B 2.1.1 Physicochemical Properties. For new biological entities (NBE) the primary structure assessment by PM on the protein level is accompanied by cDNA sequencing of the plasmid insert used for transfection. Having these two levels together creates higher safety for assurance of correct sequence especially for isobaric aminoacids (leucine, isoleucine), which are difficult to distinguish on the protein level by MS. In contrast, for the development of biosimilar drugs, no cDNA sequence is available as this is the property of the originator developers. Thus, in this case confirmation of the correct sequence depends on public amino acid sequence information on the one hand and on the other hand on analysis on the protein level. To assure the correct sequence of biosimilar candidates therefore typically several different digestions of the Originator drug are performed and in many cases not only MS is applied but also Edman degradation on individual peptides after fractionation to confirm leucine or isoleucin positions.

Furthermore, posttranslational and cotranslational modifications can be determined in a site-specific manner. These data are important for characterization as well as for comparability.

The following table lists typical modifications of individual amino acids, which are addressed during characterization and sometimes also for release testing (Table 4.2):

Analytical evaluation of these different modifications and scope of these analyses will be described in more detail in Chapter 7, Advanced Structural Elucidation of New Antibody Formats by Mass Spectrometry.

Table 4.2 Important posttranslational and cotranslational modifications occuring in mAb biotherapeutics.

Protein/amino acid	Modification
N-Terminus	Pyroglutamate formation (PyrE) Truncation Prolongation (with signal peptide)
C-Terminus	Truncation, lysine heterogeneity
Cysteine (C)	Correct intra- and intermolecular disulfide bridging Free thiols Thioethers Trisulfides Oxidation
Asparagine (N)	N-glycosylation Deamidation Succinimide formation
Glutamine (N)	Pyroglutamate formation (PyrE) Deamidation
Aspartatic acid (D)	Isomerization Succinimide formation
Serine (S), Threonine (T), and Tyrosine (Y)	O-glycosylation Hydroxylation Phosphorylation
Methionine (M) and Tryptophane (W)	Oxidation

4.2.5 Multiattribute method approach

A new analytical paradigm is to replace a broad pattern of different analytical techniques, including capillary electrophoresis and different LC methods such as RP-HPLC, ion exchange chromatography, and hydrophobic interaction chromatography by a MS-based approach, the so-called MAM (Rogers et al., 2015). The MAM approach can be applied for early development stages and characterization, for real-time monitoring of biotechnological production stability as well as for release testing and stability analysis. The advantage of the MAM approach is that with one method, that is, MS, almost all quality attributes can be addressed in a highly specific quantitative and site-specific manner. A major challenge of the approach is that the complex sample preparation has to be robust and reproducible to generate quantitative data. In a promising recent interlaboratory PM study, Millán-Martín et al. (2020) evaluated variances in four different laboratories located in four different European countries in order to get an idea about the method to be stable enough in a quality control setting. A further challenge of the MAM approach is complexity and vastness of the raw data produced. From a regulatory perspective, there are several pitfalls for handling of so many data within a pharmaceutical validation approach. However, bioinformatics, digitalization, and artificial intelligence rapidly develop further and implementation of the MAM-approach into the pharmaceutical development, production, and quality control environment steadily becomes more popular.

MS as a key technology in the biopharmaceutical industry will be now described in more detail within this Chapter as well as more detailed information of the MAM approach given in Chapter 7, Advanced Structural Elucidation of New Antibody Formats by Mass Spectrometry.

4.3 Mass spectrometry

Liquid chromatography with MS or LC-MS is the most common "hyphenated" technique. The power of LC-MS is due to the combination of chemical selectivity of analytes with a stationary phase and a mass spectrometer which can measure the mass to charge ratio of eluting peaks. In the case of PM the ability to generate a protein-specific elution pattern of digested peptides and the subsequent molecular weight information provides great depth of data linked to the peptide fingerprint. The elution pattern should be highly reproducible and is related to the specific amino acid compostition of each peptide. Shorter and more hydrophilic peptides will normally elute very early in the gradient but larger, more hydrophobic peptides tend to retain longer on the column. In addition, the ability to use MS/MS fragmentation allows the generation of data which can be used to determine the amino acid sequence of each eluting peptide. Before LC-MS/MS PM, scientists needed to collect each eluting peptide and sequence them one at a time by Edmann degradation which was very labor intensive and time consuming. The introduction of LC-MS-MS PM has enabled scientists to confirm the identity of a protein's primary structure in hours instead of weeks. Nearly any study presented in the primary literature discussing a mAb and its structure, modifications it may have undergone, its inherent heterogeneity, etc will certainly include PM with mass spec detection as a critical piece of data supporting the paper's claims. The technique easily allows for the quantitation of site-specific modifications, the presence of mutations, and relative quantitation of microheterogeneity such as that found in a glycopeptide. It is easy to see why PM using LC-MS is a critical tool for analytical scientists working with recombinant mAbs.

4.3.1 Mobile phases and consideration

PM with LC-MS detection is almost exclusively performed using reversed phase chromatography. Briefly, in a reversed phase peptide map, the digested peptides are injected onto a C18 column under aqueous conditions, typically water, and the mobile phase is then increased gradually to a largely organic composition, typically acetonitrile, which causes peptides to elute from the column based on weakening of their hydrophobic interaction. Thus the stronger the hydrophobic interaction, the longer the peptide will bind to the column. In traditional PM the eluting peaks are detected by UV absorbance at 214 nm. There are some distinct differences between this approach and PM with LC-MS detection. Typical reversed phase chromatography uses trifluoroacetic acid (TFA) as an ion-pairing agent which assists in improving the chromatographic shape of eluting peaks; however, it is not amenable to LC-MS ionization. TFA serves to mask some of the positive charges on the basic residues to allow the hydrophobicity of the peptides to be the primary factor in the reversed phase selectivity. This also increases the binding of smaller hydrophilic peptides. The problem is that the association of TFA with a peptide suppresses the ionization efficiency being the most important attribute for mass spec detection. To mitigate the problems encountered with TFA, scientists have either reduced or eliminated its use in the mobile phases. Often reversed phase mobile phases use formic acid (FA) as a mobile phase additive. The FA does not provide the chromatographic benefits of TFA but does reduce the pH to assist ionization which will be discussed later. Sometimes scientists use a blend of FA/TFA which serves as a compromise to improve peak shape without all the ionization challenges. An example of a typical blend may be 0.08% FA/0.02% TFA in each of the mobile phases.

4.3.2 Gradient and scale

One strategy to improve the data obtained for a peptide map of a mAb with mass spec detection is to increase the length of the gradient used for the separation. By making the gradient longer, the amount of time used to change the mobile phase composition from aqueous to organic increases. This allows for very subtle differences in hydrophobicity to be precisely modulated. This results in more time between closely eluting chromatographic peaks. In addition, the more homogenous the population of the analyte entering the electrospray source, the better the quality of the resulting mass spec data. It can be more challenging to discern multiple coeluting peaks, especially when some are of low and others of very high abundance. For this reason, peptide maps can be up to 3 h long per run. It can be valuable to spend some time optimizing the separation and the time of the gradient. In general, for an information rich peptide map, it is usually a good idea to put the quality of data as more important criteria rather than run time, if possible.

PM of recombinant mAbs can be performed at different chromatographic scales. The most common chromatography scales used are analytical, micro, and nano. Analytical utilizes flow rates of around 1 mL min^{-1} are used to deliver the mobile phase to the column. Typical analytical column dimensions are 4.6 mM ID with column lengths from 150 to 300 mM. For analytical flows, the efluent is usually only going to a UV detector or split between a UV detector and a mass spectrometer so as to not saturate the MS. Typical microscale flow rates are as high as 0.2 mL min^{-1} down to as low as 25 μL min^{-1}. For a 0.2 mL min^{-1} flow rate, a 2.1 mM ID column would be used. For 25–50 μL min^{-1}, a 1 mM ID column is typical. Most mass spectrometers can handle these flow rates. For naonscale separation, steel columns are no longer used. These flow rates are below 1 μL min^{-1}; therefore it is typical that a packed glass capillary column would be used. In situations when one is very sample limited, nanoscale can help to get a better signal from higher concentrations of tight bands eluting from the capillary. The scale of the chromatography is usually determined by the amount of the sample being analyzed.

4.3.3 Ionization techniques for the analysis of peptides

MS requires ionization of the analyte in order to detect it. Simply put, ionization is the process of converting an analyte into a cation by adding a proton or into an anion by removing a proton. For small molecules, the molecular structure is used to determine which of these processes is more easily achieved. For MS of proteins, producing cations is the method of choice. Peptide maps are performed using acid as an ion-pairing agent added to the mobile phases. The resulting pH of a FA or TFA/FA blended reversed phase mobile phases is about 2.5. At this pH, arginine, lysine, and histidine residues will be protonated. In addition, if trypsin is used to generate the peptides, nearly every peptide will have an arginine or lysine at its C-terminus which ensures it exists as a cation. Conversely, the primary amine at the N-terminus of each peptide can also carry a proton. For these reasons, LC-MS-MS is always performed in the positive-ion mode. The most common ionization techniques utilized for PM of proteins will be discussed in the next section.

4.3.3.1 Electrospray ionization

Electrospray ionization (ESI) is the most commonly used ionization technique for the analysis of proteins and peptides mainly due to its compatibility with reversed phase liquid chromatography. ESI was developed by Professor John Fenn whom was a corecipient of a Nobel Prize in chemistry in 2002 for this accomplishment. Electrospray takes advantage of the low pH mobile phases to induce ionization of the side chains of the basic amino acids. The technique differs from thermospray in that there is a charge stripping of negatively charged species at the spray needle which improves the cation ion transmission. A nebulizer gas generates an aerosol spray from the spray needle known as a taylor cone. Heat applied to the mass spec source causes the charged droplet to evaporate to produce a smaller droplet with a high density of positive ions. When the repulsion of the ions exceeds the surface tension of the droplet, the Raliegh limit has been met which leads to a coulombic explosion of the droplet into many tiny droplets. The process repeats until free positively charged ions pass through the sample cone to the ion optics region of the instrument. Most mass spectrometers do not spray directly into the cone in order to reduce the introduction of neutrals into the system. Electrospray can be operated in positive mode, negative mode, or with polarity switching; however, this is usually reserved for small molecule analysis and is not amenable to the analysis of proteins and peptides as mentioned earlier.

4.3.3.2 Matrix-assisted laser desorption ionization

Matrix-assisted laser desorption ionization (MALDI) is an ionization technique developed by Hillenkamp and Karras and first applied to the analysis of proteins by Koichi Tanaka, for which he was the other corecipient of the 2002 Nobel Prize in chemistry. MALDI is a technique which samples analytes from the surface of a target. For these reason, it has made a reemergence in recent years for applications such as tissue imaging.

The analysis of peptides using MALDI was conducted using a peptide map of sorts in which protein spots separated on a two dimensional gel could be excised and subjected to an in-gel digestion. The peptides could then be extracted and spotted on to a steel target. A matrix material composed of alpha-cyano-4-hydroxycinnamic acid or CHCA is typically used for peptide analysis using MALDI.

MALDI is also suitable for direct analysis of peptides spotted on a MALDI target and, in general, will provide higher quality data than that generated from an in-gel digestion. MALDI is a very quick technique which provides almost instantaneous detection of peptides however it lacks any separation technique prior to analysis. Briefly, the sample is mixed with a photo-ionizable matrix and spotted on a stainless steel target. Once dried, the target is inserted into the MALDI-TOF mass spectrometer. The target is pulsed with laser light which causes a plume. The matrix molecules become ionized; however, they transfer this charge to the accompanying peptides. Following the blast, a potential is established following a short delay. The cations are popelled into the flight tube, whereas the anions move in the opposite direction. The delay in the establishment of the potential known as delayed extraction allows the plume to clear and the dissipation of interfering neutral species.

Time of flight MS, discussed in greater detail later, is highly amenable to MALDI ionization due to its fast scan rates. Every shot of the laser will generate a population of ions sitting just above the target. As ions are generated, they can be accelerated toward the detector for mass determination almost immediately. The synergy between these two processes enables rapid laser firing and the use of rastering across the surface to generate the most representative spectra of the digested peptides.

4.3.4 Mass detectors

Mass detector refers to the mass spectrometer itself following the generation of the ions. The mass detector is the region of the mass spectrometer beyond the source which can transmit the full spectrum of all ions entering the high vacuum region and deliver them to the detection dynode for recording of their mass to charge ratios. The mass detector can also isolate individual ions for subsequent fragmentation in order to determine the amino acid sequence of that mass. A few of the more common platforms used for PM will be discussed in more detail.

4.3.4.1 Quadrupole time of flight

Quadrupole time of flight (Q-TOF) mass detectors which are capable of performing MS and MS-MS. What this means is that in the MS mode, ions are passed through the quadrupole to the entrance of the flight tube. In this region, ions are then pulsed by a pusher of like charge which sends the ions accelerating down the flight tube and into a series of reflectrons. Reflectrons are a series of rings with increasing positive charge. Eventually, the cations will stop penetrating the reflectron or ion mirror and will be repelled at an altered angle designed to transmit the cations to the ion detector. The mass/charge ratio of an ion will influence the time it takes for the ion to get from the pusher to the detector. Thus the mass can be calculated by experimentally determining the time of flight and inserting it into the equation $KE = 1/2mv^2$. The KE (kinetic energy) is the force applied by the pusher that puts the ions in motion within the flight tube at 10^{-7} torr. The velocity is equal to distance over time; therefore the time of flight is experimentally determined. The distance from the pusher to the detector is known for the system and will not change.

In the MS mode, ionized peptides are passed directly through the quadrupole, accelerated down the time of flight tube, and their mass to charge ratios are calculated as well as their relative intensities. In data-dependent acquisition (DDA), the user defines the number of ions for isolation within a spectra. For example, if the number three is used, the three most intense ions in the spectra will be selected for a sequential passage through the quadrupole region. Next, the most intense isolated ion will enter the collision cell where it will collide with gas molecules causing amino acids to be lost either from the N- or C-terminal side of the peptide. These neutral losses result in B ions, losses from the C-termini with an intact N-terminus or Y ions, losses from the N-termini with an intact C-terminus. This collection of fragments is then analyzed in the time of flight segment of the instrument generating an MS-MS spectra. Using the intact mass of the parent ion and the fragmentation pattern from the daughter ions, the amino acid sequence can be deduced. The process will then be repeated for the second most intense ion and then the third one. Following the third MS-MS scan will be a full MS scan; however, the three most intense ions from the previous full scan will not be considered for subsequent MS/MS scans but instead the next three most intense ions will be selected for subsequent isolation and fragmentation.

An advantage to Q-TOF instruments is the speed of data acquisition. The quadrupole in MS/MS mode can mass select rapidly and the TOF analyzer is also very fast. Ions are sampled continuously in either full scan or MS/MS mode. Additionally, the TOF has a wide mass dynamic range within spectra. The implications of this are that a full fragmentation pattern can be measured down to Y_1 or B_1 fragments. This is a limitation with ion trap MS/MS activation which will be discussed later. The high speed and high dynamic range means that more quality MS/MS spectra can be obtained. The results are high sequence coverage or higher identifications for proteomics approaches.

4.3.4.2 Linear ion trap

Ion trap instruments once dominated the PM landscape. They scan extremely fast and have the unique ability to perform MSn which means that a fragment of a daughter ion can be trapped and subjected to another round of MS/MS albeit with sensitivity losses at each step. Ion trap instruments utilize automatic gain control (AGC) to move ions into the trap. Briefly, a very rapid ion sample is dumped directly to the detector to measure the analyte in the electrosprayed eluent. Once determined, the instrument will trap ions for a duration defined by the user for consistent signal. It is important to note that once this level is achieved, the gate closes and additional ions from the source are being lost while the trap is in full scan or MS/MS mode.

The two major limitations of a linear ion trap are lower resolving power and the one-third rule. The mass error for an MS/MS-fragment can approach 1 Da which is less than ideal and could provide false identifications. The bigger concern is the one-third rule limitation which simply stated is the loss of the bottom one-third of the MS/MS spectra due to the inability of the trap to keep these ions in stable resonance. To address some of these limitations, linear ion traps have been coupled to higher resolving orbitrap detectors which will be discussed later, and the integration of a traditional collision cell which can retain the lower *m/z* daughter ions. In short, the linear ion trap on its own is an economical choice for performing PM detection at a high sampling rate if the limitations can be tolerated.

4.3.4.3 Orbitrap platforms

The orbitrap technology was invented by Alexander Makarov. He was able to simulate the orbital movement of ions typical for ultra-high resolving FT-ICR instruments without the long scan times and extreme temperatures required to cool the high-field magnets. Dr. Makarov was able to simulate the ion motion using electric fields. The initial instruments were hybrids in which a linear ion trap was coupled to an orbitrap analyzer. Following the AGC scan, the ions selected by the linear ion trap were transferred and held in the C-trap. Next, the ions were transferred to the orbitrap where they circled the center ring electrode and were spatially resolved based on their m/z value and then subsequently detected. The orbitrap overcomes the low-resolving power of the linear ion trap. However, it does so at the expense of speed. The orbitrap can provide superior resolution to a TOF analyzer but cannot match its scan rate.

The second-generation orbitrap had improved resolution and scan rates and was also incorporated into new architectures, namely the Q-Exactive Plus and the Orbitrap Fusion. The Q-Exactive Plus was a hybrid instrument which mirrored the Q-TOF insstrument except replaced the TOF region with an Orbitrap. In this way, the traditional collision cell ensured that there would be no one-third rule MS/MS limitation. The Q-Exactive Plus has become one of the most widely used LC-MS platforms for performing PM. The Orbitrap Fusion incorporated this functionality along with the incorporation of a dual-stage linear ion trap.

4.3.5 Sequence analysis

The peptide map is often referred to as the protein fingerprint. The composition and the elution pattern for a given set of peptides generated from a protein is highly unique. Consequently, if one peptide map has a different elution profile than another one, even with only a single peak different, then there has been some sort of a change to the primary structure. For these reasons, the peptide map is a critical tool for the analysis of recombinant mAbs. For a given mAb, there will be chromatographic uniqueness despite the shared conserved primary structure accross this class of molecules. It is important to not overlook the value of the mAb chromatographic fingerprint. For LC-MS-MS PM of mAbs, it is often the case that the eluent may first pass through a UV detector and the travel to the mass spec elctrospray source. In addition to the UV chromatogram, the elution pattern of the full scan can be plotted as the total ion current (TIC). Peptide maps can be stacked and carefully compared for any differences. It is common to try to match the identity of peptides in the UV chromatogram with those in the TIC where the MS and MS/MS data can be used to determine amino acid sequence.

4.3.5.1 Manual sequence verification

The most hands-on approach to PM with MS for sequence verification of a recombinant mAb is a manual approach. Manual approaches typically require a software application which can generate an in silico peptide list and a coresponding fragment ion list for each peptide. The first step to the manual approach is to extract the mass trace for a corresponding *m/z* for the peptide being searched. This is typically singly charged for low-mass peptides, doubly charged for peptides with molecular weights between 500 and 1000 Da and correspondingly greater charge states as the mass increases. For a high-quality extracted ion chromatogram (EIC), there is typically a single peak which dominates the trace. An extraction of the spectra should reveal the corresponding mass(es) of the peptide charge state(s) in the full spectrum. This confirms the peptide mass. Confirming the sequence from the MS/MS spectra requires that we "walk" through the daughter ion scans looking for the scan which corresponds to one of the peptide charge states. Once found, we confirm the observed b-ion and y-ion fragments from our in silico fragment list. The unambiguous confirmation of the peptide ID is achieved by confirming ions from both series which ovelap the peptide sequence. Now that this peptide has been confirmed, and the next ion in the in silico peptide list is searched in the same manner. It is important to point out two important considerations. First, some peptides may exist almost solely with a posttranslational modification which should be accounted for in both of the above steps. For example, a protein with an N-terminal glutamine will cyclize into pyroglutamate which reduces the mass by 17 daltons. This minus 17 Da shift must be corrected for both the EIC generation and the observed b-ion fragments. Second, some small early eluting peptides may have poor ionization complicating their detection, EIC generation, and MS/MS selection. For this reason, a multiple enzyme approach will be discussed next.

4.3.5.2 Multiple proteases approach

As discussed in the previous section, sequence verification of small, hydrophilic peptides can be challenging. The use of multiple proteases in the analysis of a recombinant mAb can effectively address these challenges. Trypsin is the most universal protease used in LC-MS-MS PM workflows because of its cost, efficiency, and specificity. Virtually all tryptic peptides will have a lysine or an arginine at its N-terminus which serves to enhance ionization; however, as already mentioned, small peptides may still go unaccounted for in the analysis. The use of other proteases such as Glu-C, Asp-N, or Chymotrypsin provides very different cleavage specificity. Using a comprehensive multiple digest approach involves the preparation of the same in four different tubes. Each tube is treated with a different protease, and the analysis is conducted by running four LC-MS-MS peptide maps, each treated with a different enzyme. The in silico lists and data analysis are conducted the same way as described in the previous section. In this way, it is almost certain that the entire primary structure of the antibody will be covered by high confidence MS and MS/MS data.

The following hypothetical peptide fragments illustrate how the use of multple enzymes can help to deduce what the amino acid sequence is in a protein:

Trypsin	QEDR/HAAWSETVK/GSY
LysC	QEDRHAAWSETVK/GSY
Chymo	QEDRHAAW/SETVKGSY
Glu-C	QED/RHAAWSE/TVKGSY
Asp-N	QE/DRHAAWSETVKGSY

It is very easilly seen how the use of multiple enzymes can assist in producing complimentary peptide which can make the identification of the amino acid sequence more successful. It is not always necessary to use all of these different proteases, and one can save time and money by performing the digests which produce the most relevant peptides to underrepresented regions of a protein using standard tryptic mapping.

Automated approaches using a FASTA file: As biopharma business models evolved toward the use of automation for upstream and downstream development, the number of mAb samples for sequence analysis has grown significantly. The manual approach, while very thorough, is very time consuming and not feasible for large numbers of samples. To address this, softwares have been developed which can automatically determine the sequence coverage and confirmation in a peptide map. Some of these software applications are open access but the most well known are either supplied by the mass spec vendor or are "vendor neutral" subscription-based packages. To perform these workflows, the user must enter the FASTA sequence into the software using the following format:

> Protein Name
THEPRTEINSEQ (in one letter code)

In addition to the FASTA sequence, the user must define the protease used, the alkylating reagent, posttranslational modifications, and target residues that may be present, the number of miscleavages to be considered and the expected peptide mass ranges. Most software applications used for sequence verification use MS/MS data to determine the peptides which are present as well as any PTMs. The output file will also report the sequence coverage percentage confirmed. An attractive function of these automated approaches is the identification of all confirmed peptides on the FASTA sequence which allows for a quick visual assessment of the primary structure of the protein.

4.3.5.3 Sequence variant analysis

The detection of sequence variants is usually done by using complementary analytical techniques along with MS for confirmation. To this point, often a sequence variant is suspected when the intact or reduced molecular weights do not match the expected theoretical molecular weight. In addition, often a change in the charge profile by weak cation exchange chromatography or icIEF is observed. In simple cases, a mass deviation may be assigned to a specific amino acid substitution which can then be evaluated at the nucleotide level. If the substitution is the result of a single-nucleotide change, then it is possible and is worth digging into the data to determine if it is truly present.

The sequence variant can be located by evaluating the peptides in a manual manner; however, this is very labor intensive. Using a computational tool with either a suspected misincorporation or a "undefined" search can often provide a positive hit for a misincorporation. A manual evaluation of the MS/MS data as described earlier can conclusively identify the mutation position. Software tools have been developed which build lists of potential sequence variants. It is then up to the user to go through the list to evaluate whether it is real or a false positive. The nature of this tool is to be over cautious in its identification of sequence variants; therefore the vast majority of the hits are false positives. To determine whether a hit is real or not, the user must compare the ms spectra between the WT and variant peptides, along with the MS/MS data quality and spectral similarities, and lastly, whether the mutation makes sense. In addition to a reasonable nucleotide change leading to the primary structure difference, the chromatographic behavior of the mutant peptide should make sense. A mutation of a hydrophilic residue such as serine to a very hydrophobic residue such as leucine should result in a longer retention time. If this is not observed, the mutation should be questioned. In addition, it if the peptide map suggests large amounts of a sequence variant which is not supported by other middle down approaches, the variant may not be real. Sequnce variant analysis should be performed in a very holistic manner (Li, Wypych, & Duff, 2017).

4.3.5.4 De novo sequencing

For instances where a primary structure is not known for a protein, *de novo* sequencing is used to determine peptide amino acid sequences. This approach is very effective in helping to deduce the amino acid sequences of a mAb's CDR regions which may not be known. *De novo* sequencing is a tool which evaluates the peptide *m/z* and the neutral loss pattern from the spectra to deduce the amino acid sequence when it is not known. Most *de novo* sequencing is performed using automated software packages. The putative hits are then manually examined and adjusted by the user if a better interpretations can be made. Additionally, these software tools perform primary structure alignments against database looking for molecules with high homology in order to make higher probability suggestions based on framework sequences within that protein class. *De novo* sequencing is likely the most challenging application of LC-MS-MS PM (O'Bryon, Jenson, & Merkley, 2020).

4.4 Conclusion

Analytical assessment of primary structure of mAb products is a basic requirement for approval of NBE candidates as well as for novel mAb-based biosimilar candidates. Traditional analytical techniques such as automated Edman degradation are today only used to answer special analytical questions such as distinguishing isobaric aminoacids such as leucine and isoleucine. PM with analytical readout by UV absorbtion and in particular by MS has become the gold standard in industry for unambiguous assessment of correct primary structure. Furthermore, MS is capable to specifically and quantitatively determine cotranslational (glycosylation) and posttranslational modifications such as oxidation of methione and tryptophane, deamidation of asparagine and glutamine, isomerization of aspartatic acid, formation of pyro-glutamate from N-terminal glutamine as well as truncations at either termini. Therefore, MS is an essential analytical tool throughout the biopharmaceutical value chain including developability assessment, analytical process support, characterization and comparability, and finally for release and stability studies. Rapid development during the last years regarding MS instrumentation as well as of bioinformatic tools made MS more and more attractive for the biopharmaceutical industry. These developments are basis for an analytical paradigm shift to moving from a comprehensive set of traditional analytical methods based on liquid chromatography and electrophoretic separation more towards using MS as MAM as almost universal tool.

References

Baker, J. J., McDaniel, D., Cain, D., Lee Tao, P., Li, C., Huang, Y., ... Niñonuevo, M. (2019). Rapid identification of disulfide bonds and cysteine-related variants in an IgG1 knob-into-hole bispecific antibody enhanced by machine learning. *Analytical Chemistry*, *91*, 965–976.

Cheng, Y., Chen, M. T., Patterson, L. C., Yu, X. C., Zhang, Y. T., Burgess, B. L., & Chen, Y. (2017). Domain-specific free thiol variant characterization of an IgG1 by reversed-phase high-performance liquid chromatography mass spectrometry. *Analytical Biochemistry*, *519*, 8–14.

Crimmins, D. L., Mische, S. M., & Denslow, N. D. (2005). Chemical cleavage of proteins in solution. *Current Protocols in Protein Science*, 1104–1140.

Dong, M. W. (1992). Tryptic mapping by reversed phase liquid chromatography. *Advances in Chromatography*, *32*, 21–51.

Dong, M. W., & Tran, A. D. (1990). Factors influencing the performance of peptide mapping by reversed-phase high-performance liquid chromatography. *Journal of Chromatography*, *499*, 125–139.

Dougherty, J. J., Jr, Snyder, L. M., Sinclair, R. L., & Robins, R. H. (1990). High-performance tryptic mapping of recombinant bovine somatotropin. *Analytical Biochemistry*, *190*, 7–20.

Fruton, J. S. (1992). An episode in the history of protein chemistry: Pehr Edman's method for the sequential degradation of peptides. *International Journal of Peptide and Protein Research*, *39*, 189–194.

Jones, M. D., Merewether, L. A., Clogston, C. L., & Lu, H. S. (1994). Peptide map analysis of recombinant human granulocyte colony stimulating factor: elimination of methionine modification and nonspecific cleavages. *Analytical Biochemistry*, *216*, 135–146.

Kannan, K., Mulkerrin, M. G., Zhang, M., Gray, R., Steinharter, T., Sewerin, K., ... Karunatilake, C. (1997). Rapid analytical tryptic mapping of a recombinant chimeric monoclonal antibody and method validation challenges. *Journal of Pharmaceutical and Biomedical Analysis*, *16*, 631–640.

Kita, A., Ponniah, G., Nowak, C., & Liu, H. (2016). Characterization of cysteinylation and trisulfide bonds in a recombinant monoclonal antibody. *Analytical Chemistry*, *88*, 5430–5437.

Kori, Y., Patel, R., Neill, A., & Liu, H. (2015). A conventional procedure to reduce Asn deamidation artifacts during trypsin peptide mapping. *Journal of Chromatography. B, Analytical Technologies in the Biomedical and Life Sciences*, *2015*, 107–113.

Li, W., Wypych, J., & Duff, R. J. (2017). Improved sequence variant analysis strategy by automated false positive removal. *MAbs*, *9*(6), 978–984.

Liu, H., Gaza-Bulseco, G., & Sun, J. (2006). Characterization of the stability of a fully human monoclonal IgG after prolonged incubation at elevated temperature. *Journal of Chromatography. B, Analytical Technologies in the Biomedical and Life Sciences*, *837*, 35–43.

Medzihradszky, K. F. (2008). Characterization of site-specific N-glycosylation. *Methods in Molecular Biology*, *446*, 293–316.

Millán-Martín, S., Jakes, C., Carillo, S., Buchanan, T., Guender, M., Kristensen, D. B., ... Bones, J. (2020). Inter-laboratory study of an optimised peptide mapping workflow using automated trypsin digestion for monitoring monoclonal antibody product quality attributes. *Analytical and Bioanalytical Chemistry*, *412*, 6833–6848.

Mouchahoir, T., & Schiel, J. E. (2018). Development of an LC-MS/MS peptide mapping protocol for the NISTmAb. *Analytical and Bioanalytical Chemistry*, *410*(8), 2111–2126.

O'Bryon, I., Jenson, S. C., & Merkley, E. D. (2020). Flying blind, or just flying under the radar? The underappreciated power of *de novo* methods of mass spectrometric peptide identification. *Protein Science*, *29*, 1864–1878.

Pace, A. L., Wong, R. L., Zhang, Y. T., Kao, Y. H., & Wang, Y. J. (2013). Asparagine deamidation dependence on buffer type, pH, and temperature. *Journal of Pharmaceutical Sciences*, *102*, 1712–1723.

Robinson, A. B., Scotchler, J. W., & McKerrow, J. H. (1973). Rates of nonenzymatic deamidation of glutaminyl and asparaginyl residues in pentapeptides. *Journal of the American Chemical Society*, *95*, 8156–8159.

Rogers, R. S., Nightlinger, N. S., Livingston, B., Campbell, P., Bailey, R., & Balland, A. (2015). Development of a quantitative mass spectrometry multi-attribute method for characterization, quality control testing and disposition of biologics. *mAbs*, *7*, 881–890.

Schechter, I., & Berger, A. (1968). On the active site of proteases. Mapping the active site of papain; specific peptide inhibitors of papain. *Biochemical and Biophysical Research Communications*, *32*, 898–902.

Swiss Institute of Bioinformatics Resource Portal Expasy, PeptideCutter Tool: https://web.expasy.org/peptide_cutter/peptidecutter_enzymes.html

Tassi, M., De Vos, J., Chatterjee, S., Sobott, F., Bones, J., & Eeltink, S. (2018). Advances in native high performance liquid chromatography and intact mass spectrometry for the characterization of biopharmaceutical products. *Journal of Separation Science*, *41*, 125–144.

Further reading

Allen, D., Baffi, R., Bausch, J., Bongers, J., Costello, M., Dougherty, J., Jr, ... Tuls, J. (1996). Validation of peptide mapping for protein identity and genetic stability. Biologics and biotechnology section, pharmaceutical research and manufacturers of America. *Biologicals*, *24*, 255–274.

Allen, J. S. (1947). An improved electron multiplier particle counter. *Review of Scientific Instruments*, *18*(10), 739–749.

Amster, I. J. (1996). Fourier transform mass spectrometry. *Journal of Mass Spectrometry*, *31* (12), 1325–1337.

Anacleto, J. F., Pleasance, S., & Boyd, R. K. (1992). Calibration of ion spray mass spectra using cluster ions. *Organic Mass Spectrometry.*, *27*, 660–666.

Baily, A. G. (1988). *Electrostatic spraying of liquids*. New York, NY, USA: John Wiley & Sons.

Banks, J. F., Quinn, J. P., & Whilehouse, C. M. (1994). LC/ESI-MS determination of proteins using conventional liquid chromatography and ultrasonically assisted electrospray. *Analytical Chemistry*, *66*(21), 3688–3695.

Banks, J. F., Shen, S., Whitehouse, C. M., & Fenn, J. B. (1994). Ultrasonically assisted electrospray ionization for LC/MS determination of nucleosides from a transfer RNA digest. *Analytical Chemistry*, *66*(3), 406–414.

Barber, M., Bordoli, R. S., Sedgwick, R. D., & Tyler, A. N. (1981). Fast atom bombardment of solids (F.A.B.): a new ion source for mass spectrometry. *Journal of the Chemical Society, Chemical Communications* (7), 325–327.

Bereman, M. S., Nyadong, L., Fernandez, F. M., & Muddiman, D. C. (2006). Direct high-resolution peptide and protein analysis by desorption electrospray ionization Fourier transform ion cyclotron resonance mass spectrometry. *Rapid Communications in Mass Spectrometry*, *20*, 3409–3411.

Berquist, J., Palmblad, M., Wetterhall, M., Hakansson, P., & Markides, K. (2002). Peptide mapping of proteins in human body fluids using electrospray ionization Fourier transform ion cyclotron resonance mass spectrometry. *Mass Spectrometry Reviews*, *21*(1), 2–15.

Bian, Y., Zheng, R., Bayer, F. P., et al. (2020). Robust, reproducible and quantitative analysis of thousands of proteomes by micro-flow LC−MS/MS. *Nature Communications*, *11*, 157.
Biemann, K. (1990). Sequencing of peptides by tandem mass spectrometry and high-energy collision-induced dissociation. *Methods in Enzymology*, *193*, 455−479.
Biemann, K., & Scoble, H. A. (1987). Characterization by tandem mass spectrometry of structural modifications in proteins. *Science*, *237*, 992−998.
Biewer, T., Alexander, D., Robertson, S., & Walch, B. (1994). Electrostatic orrery for celestial mechanics. *American Journal of Physics*, *62*, 821−828.
Blades, A. T., Ikonomou, M. G., & Kebarle, P. (1991). Mechanism of electrospray mass spectrometry. Electrospray as an electrolysis cell. *Analytical Chemistry*, *63*(19), 2109−2114.
Bruins, A. P. (1991). Mass spectrometry with ion sources operating at atmospheric pressure. *Mass Spectrometry Reviews*, *10*(1), 53−77.
Bruins, A. P., Covey, T. R., & Henion, J. D. (1987). Ion spray interface for combined liquid chromatography/atmospheric pressure ionization mass spectrometry. *Analytical Chemistry*, *59*(22), 2642−2646.
Cech, N. B., & Enke, C. G. (2001). Practical implications of some recent studies in electrospray ionization fundamentals. *Mass Spectrometry Reviews*, *20*(6), 362−387.
Chapman, J. R., Gallagher, R. T., Barton, E. C., Curtis, J. M., & Derrick, P. J. (1992). Advantages of high-resolution and high-mass range magnetic-sector mass spectrometry for electrospray ionization. *Organic Mass Spectrometry.*, *27*, 195−203.
Chaurand, P., Luetzenkirchen, F., & Spengler, B. (1999). Peptide and protein identification by matrix-assisted laser desorption ionization (MALDI) and MALDI-post-source decay time-of-flight mass spectrometry. *Journal of the American Society for Mass Spectrometry*, *10*, 91−103.
Chen, L., & Marshall, A. G. (1987). Stored waveform simultaneous mass-selective ejection/excitation for Fourier transform ion cyclotron resonance mass spectrometry. *International Journal of Mass Spectrometry and Ion Processes*, *79*, 115−125.
Chernushevich, I. V., Loboda, A. V., & Thomson, B. A. (2001). An Introduction to quadrupole-time-of-flight mass spectrometry. *Journal of Mass Spectrometry*, *36*, 849−865.
Cloupeau, M., & Prunet-Foch, B. (1990). Electrostatic spraying of liquids: Main functioning. *Journal of Electrostatics*, *25*, 165−184.
Cole, R. B. (2000). Some tenets pertaining to electrospray ionization mass spectrometry. *Journal of Mass Spectrometry*, *35*(7), 763−772.
Cole, R. B. (2010). *Electrospray and MALDI mass spectrometry*. New Jersey, NJ, USA: John Wiley & Sons.
Cooks, R. G., Cleven, C. D., Horn, L. A., Nappi, M., Weil, C., Soni, M. H., & Julian, R. K. J. (1995). Non-destructive detection of ions in a quadrupole ion trap using a d. c. pulse to force coherent ion motion: A simulation study. *International Journal of Mass Spectrometry and Ion Processes*, *146/147*, 147−163.
Covey, T. R., Bonner, R. F., Shushan, B. I., & Henion, J. (1988). The determination of protein, oligonucleotide and peptide molecular weights by ion-spray mass spectrometry. *Rapid Communications in Mass Spectrometry*, *2*(11), 249−256.
Covey, T. R., Bruins, A. P., & Henion, J. D. (1988). Comparison of thermospray and ion spray mass spectrometry in an atmospheric pressure ion source. *Organic Mass Spectrometry.*, *23*, 178−186.
Cox, K. A., Cleven, C. D., & Cooks, R. G. (1995). Mass shifts and local space charge effects observed in the quadrupole ion trap at higher resolution. *International Journal of Mass Spectrometry and Ion Processes*, *144*, 47−65.
Diehl, G., & Karst, U. (2002). On-line electrochemistry—MS and related techniques. *Analytical and Bioanalytical Chemistry*, *373*(6), 390−398.

Dole, M., Mack, L. L., Hines, R. L., et al. (1968). Molecular beams of macroions. *The Journal of Chemical Physics*, *49*(5), 2240–2249.

Domon, B., & Aebersold, R. (2006). Mass spectrometry and protein analysis. *Science*, *312* (5771), 212–217.

El-Aneed, A., Cohen, A., & Banoub, J. (2009). Mass spectrometry, review of the basics: Electrospray, MALDI, and commonly used mass analyzers. *Applied Spectroscopy Reviews*, *44*(3), 210–230.

European Medicines Agency. (1999). ICH Topic Q6B. Specifications: Test Procedures and Acceptance Criteria for Biotechnological/Biological Products https://www.ema.europa.eu/en/documents/scientific-guideline/ich-q-6-b-test-procedures-acceptance-criteria-biotechnological/biological-products-step-5_en.pdf

European Medicines Agency. (2005). ICH Topic Q5E. Comparability of Biotechnological/Biological Products. https://www.ema.europa.eu/en/documents/scientific-guideline/ich-q-5-e-comparability-biotechnological/biological-products-step-5_en.pdf

Faulstich, K., Worner, K., Brill, H., & Engels, J. W. (1997). A sequencing method for RNA oligonucleotides based on mass spectrometry. *Analytical Chemistry*, *69*, 4349–4353.

Fenn, J. B. (1993). Ion formation from charged droplets: Roles of geometry, energy, and time. *Journal of the American Society for Mass Spectrometry*, *4*(7), 524–535.

Fenn, J. B. (2003). Electrospray wings for molecular elephants (Nobel lecture). *Angewandte Chemie*, *42*(33), 3871–3894.

Fenn, J., Mann, M., Meng, C. K., Wong, S. F., & Whitehouse, C. (1990). Electrospray ionization-principles and practice. *Mass Spectrometry Reviews*, *9*(1), 37–70.

Fenn, J. B., Mann, M., Meng, C. K., Wong, S. F., & Whitehouse, C. M. (1989). Electrospray ionization for mass spectrometry of large biomolecules. *Science*, *246* (4926), 64–71.

García, M. C. (2005). The effect of the mobile phase additives on sensitivity in the analysis of peptides and proteins by high-performance liquid chromatography-electrospray mass spectrometry. *Journal of Chromatography. B, Analytical Technologies in the Biomedical and Life Sciences*, *825*(2), 111–123.

Gillig, K. J., Bluhm, B. K., & Russell, D. H. (1996). Ion motion in a Fourier transform ion cyclotron resonance wire ion guide cell. *International Journal of Mass Spectrometry*, *157/158*, 129–147.

Goeringer, D. E., Crutcher, R. I., & McLuckey, S. A. (1995). Ion remeasurement in the radio frequency quadrupole ion trap. *Analytical Chemistry*, *67*, 4164–4169.

Gologan, B., Takáts, Z., Alvarez, J., et al. (2004). Ion soft-landing into liquids: Protein identification, separation, and purification with retention of biological activity. *Journal of the American Society for Mass Spectrometry*, *15*(12), 1874–1884.

Grayson, M. (2011). John Bennett Fenn: A curious road to the prize. *Journal of the American Society for Mass Spectrometry*, *22*(8), 1301–1308.

Gross, J. H. (2004). *Mass spectrometry*. Heidelberg, Germany: Springer.

Guevremont, R., Siu, K. W. M., Le Blanc, J. C. Y., & Berman, S. S. (1992). Are the electrospray mass spectra of proteins related to their aqueous solution chemistry? *Journal of the American Society for Mass Spectrometry*, *3*(3), 216–224.

Hager, J. W. (2002). A new linear ion trap mass spectrometer. *Rapid Communications in Mass Spectrometry*, *16*, 512–526.

Hahne, H., et al. (2013). DMSO enhances electrospray response, boosting sensitivity of proteomic experiments. *Nature Methods*, *10*, 989.

Hoff, E. R., & Chloupek, R. C. (1996). Analytical peptide mapping of recombinant DNA-derived proteins by reversed-phase high-performance liquid chromatography. *Methods in Enzymology*, *271*, 51–68.

Hoffmann, E. D., & Stroobant, V. (2001). *Mass spectrometry principles and applications* (2nd edition). Chichester, UK: John Wiley & Sons.

Hop, C. E. C. A. (1996). Generation of high molecular weight cluster ions by electrospray ionization; implications for mass calibration. *Journal of Mass Spectrometry*, *31*, 1314–1316.

Hop, C. E. C. A., & Bakhtiar, R. (1996). Electrospray ionization mass spectrometry. Part III: Applications in inorganic chemistry and synthetic polymer chemistry. *Journal of Chemical Education*, *73*(8), A162–A169.

Horning, E. C., Carroll, D. I., Dzidic, I., Haegele, K. D., Horning, M. G., & Stillwell, R. N. (1974a). Atmospheric-pressure ionization mass-spectrometry—Solvent-mediated ionization of samples introduced in solution and in a liquid chromatograph effluent stream. *Journal of Chromatographic Science*, *12*, 725–729.

Houde, D., Kauppinen, P., Mhatre, R., & Lyubarskaya, Y. (2006). Determination of protein oxidation by mass spectrometry and method transfer to quality control. *Journal of Chromatography A*, *1123*(2), 189–198.

Hu, Q., Noll, R. J., Li, H. Y., Makarov, A., Hardman, M., & Cooks, R. G. (2005). The orbitrap: A new mass spectrometer. *Journal of Mass Spectrometry*, *40*, 430–443.

Hunt, D. F., Yates, J. R., Shabanowitz, J., Winston, S., & Hauer, C. R. (1986). Protein sequencing by tandem mass spectrometry. *Proceedings of the National Academy of Sciences of the United States of America*, *83*, 6233–6237.

Ikonomou, M. G., Blades, A. T., & Kebarle, P. (1991). Electrospray-ion spray: A comparison of mechanisms and performance. *Analytical Chemistry*, *63*(18), 1989–1998.

Jocelyn, P. C. (1987). Chemical reduction of disulfides. *Methods in Enzymology*, *143*, 246–256.

Karas, M., Bahr, U., & Dülcks, T. (2000). Nano-electrospray ionization mass spectrometry: Addressing analytical problems beyond routine. *Fresenius' Journal of Analytical Chemistry*, *366*(6–7), 669–676.

Karas, M., & Hillenkamp, F. (1988). Laser desorption ionization of proteins with molecular masses exceeding 10,000 Daltons. *Analytical Chemistry*, *60*, 2299–2301.

Kebarle, P., & Peschke, M. (2000). On the mechanisms by which the charged droplets produced by electrospray lead to gas phase ions. *Analytica Chimica Acta*, *406*(1), 11–35.

Kebarle, P., & Tang, L. (1993). From ions in solution to ions in the gas phase: The mechanism of electrospray mass spectrometry. *Analytical Chemistry*, *65*(22), 972A–986A.

Keith-Roach, M. J. (2010). A review of recent trends in electrospray ionisation-mass spectrometry for the analysis of metal-organic ligand complexes. *Analytica Chimica Acta*, *678*(2), 140–148.

Kelleher, N. L., Lin, H. Y., Valaskovic, G. A., Aaserud, D. J., Fridriksson, E. K., & McLafferty, F. W. (1999). Top down versus bottom up protein characterization by tandem high-resolution mass spectrometry. *Journal of the American Chemical Society*, *121*, 806–812.

Kelly, M. A., Vestling, M. M., Fenselau, C. C., & Smith, P. B. (1992). Electrospray analysis of proteins: A comparison of positive-ion and negative-ion mass spectra at high and low pH. *Organic Mass Spectrometry*, *27*, 1143–1147.

Labowsky, M., Whitehouse, C., & Fenn, J. B. (1993). Three-dimensional deconvolution of multiply charged spectra. *Rapid Communications in Mass Spectrometry*, *7*, 71–84.

Lauber, M., Koza, S., McCall, S., Alden, B., Iraneta, P., & Fountain, K. (2013). High-resolution peptide mapping separations with MS-friendly mobile phases and charge-surface-modified C18. *Analytical Chemistry*, *85*(14), 6936–6944.

Le Blanc, J. C. Y., Wang, J., Guevremont, R., & Siu, K. W. M. (1994). Electrospray mass spectra of protein cations formed in basic solutions. *Organic Mass Spectrometry*, *29*, 587–593.

Leito, I., Herodes, K., Huopolainen, M., et al. (2008). Towards the electrospray ionization mass spectrometry ionization efficiency scale of organic compounds. *Rapid Communications in Mass Spectrometry*, *22*(3), 379–384.

Li, W., Yang, B., Zhou, D., Xu, J., Ke, Z., & Suen, W. C. (2016). Discovery and characterization of antibody variants using mass spectrometry-based comparative analysis for biosimilar candidates of monoclonal antibody drugs. *Journal of Chromatography. B, Analytical Technologies in the Biomedical and Life Sciences*, *1025*, 57–67.

Li, Y., Fu, T., Liu, T., Guo, H., Guo, Q., Xu, J., et al. (2016). Characterization of alanine to valine sequence variants in the Fc region of nivolumab biosimilar produced in Chinese hamster ovary cells. *mAbs*, *8*(5), 951–960.

Ljigand, P., Kaupmees, K., & Kruve, A. (2019). Influence of the amino acid composition on the ionization efficiencies of small peptides. *Journal of Mass Spectrometry*, *54*(6), 481–487.

Loo, J. A., Edmonds, C. G., Udseth, H. R., & Smith, R. D. (1990). Effect of reducing disulfide-containing proteins on electrospray ionization mass spectra. *Analytical Chemistry*, *62*(7), 693–698.

Lundell, N., & Schreitmuller, T. (1999). Sample preparation for peptide mapping—A pharmaceutical quality-control perspective. *Analytical Biochemistry*, *266*(1), 31–47.

Macek, B., Waanders, L. F., Olsen, J. V., & Mann, M. (2006). Top-down protein sequencing and MS^3 on a hybrid linear quadrupole ion trap-orbitrap mass spectrometer. *Molecular & Cellular Proteomics*, *5*, 949–958.

Mack, L. L., Kralik, P., Rheude, A., & Dole, M. (1970). Molecular beams of macroions. II. *The Journal of Chemical Physics*, *52*(10), 4977–4986.

Mann, M., Meng, C. K., & Fenn, J. B. (1989). Interpreting mass spectra of multiply charged ions. *Analytical Chemistry*, *61*(15), 1702–1708.

March, R. E. (1997). An introduction to quadrupole ion trap mass spectrometry. *Journal of Mass Spectrometry*, *32*(4), 351–369.

McLuckey, S. A., Goeringer, D. E., & Glish, G. L. (1991). Selective ion isolation/rejection over a broad mass range in the quadrupole ion trap. *Journal of the American Society for Mass Spectrometry*, *2*, 11–21.

Oss, M., Kruve, A., Herodes, K., & Leito, I. (2010). Electrospray ionization efficiency scale of organic compound. *Analytical Chemistry*, *82*(7), 2865–2872.

Ouyang, Z., Takáts, Z., Blake, T. A., et al. (2003). Preparing protein microarrays by soft-landing of mass-selected ions. *Science*, *301*(5638), 1351–1354.

Paulovich, A. G., Billheimer, D., Ham, A. J., Vega-Montoto, L., Rudnick, P. A., Tabb, D. L., et al. (2010). Interlaboratory study characterizing a yeast performance standard for benchmarking LC-MS platform performance. *Molecular & Cellular Proteomics*, *9*(2), 242–254.

Przybylski, M., & Glocker, M. O. (1996). Electrospray mass spectrometry of biomacromolecular complexes with noncovalent interactions—New analytical perspectives for supramolecular chemistry and molecular recognition processes. *Angewandte Chemie*, *35* (8), 807–826.

Rathore, D., Faustino, A., Schiel, J., Pang, E., Boyne, M., & Rogstad, S. (2018). The role of mass spectrometry in the characterization of biologic protein products. *Expert Review of Proteomics*, *15*, 431–449.

Rudnick, P. A., Clauser, K. R., Kilpatrick, L. E., Tchekhovskoi, D. V., Neta, P., Blonder, N., et al. (2010). Performance metrics for liquid chromatography-tandem mass spectrometry systems in proteomics analyses. *Molecular & Cellular Proteomics*, *9*(2), 225–241.

Schneiderheinze, J., Walden, Z., Dufield, R., & Demarest, C. (2009). Rapid online proteolytic mapping of PEGylated rhGH for identity confirmation, quantitation of methionine oxidation and quantitation of UnPEGylated N-terminus using HPLC with UV detection. *Journal of Chromatography. B, Analytical Technologies in the Biomedical and Life Sciences*, *877*(31), 4065–4070.

Shevchenko, A., Chernushevich, I., Ens, W., Standing, K. G., Thomson, B., Wilm, M., & Mann, M. (1997). Rapid 'de Novo' peptide sequencing by a combination of nanoelectrospray, isotopic labeling and a quadrupole/time-of-flight mass spectrometer. *Rapid Communications in Mass Spectrometry*, *11*, 1015–1024.

Smith, R. D., Loo, J. A., Ogorzalek Loo, R. R., Busman, M., & Udseth, H. R. (1991). Principles and practice of electrospray ionization-mass spectrometry for large polypeptides and proteins. *Mass Spectrometry Reviews*, *10*(5), 359–451.

Smythe, C. (1936). The reaction of iodoacetate and of iodoacetamide with various sulfhydryl groups, with urease, and with yeast preparations. *The Journal of Biological Chemistry*, *114*, 601–612.

Song, Y., Schowen, R. L., Borchardt, R. T., & Topp, E. M. (2001). Effect of 'pH' on the rate of asparagine deamidation in polymeric formulations: 'pH'-rate profile. *Journal of Pharmaceutical Sciences*, *90*(2), 141–156.

Stanton, H. E., Chupka, W. A., & Inghram, M. G. (1956). Electron multipliers in mass spectrometry; effect of molecular structure. *Review of Scientific Instruments*, *27*(2), 109.

Takáts, Z., Wiseman, J. M., Gologan, B., & Cooks, R. G. (2004). Electrosonic spray ionization. A gentle technique for generating folded proteins and protein complexes in the gas phase and for studying ion-molecule reactions at atmospheric pressure. *Analytical Chemistry*, *76*(14), 4050–4058.

Thomson, J. J. (1913). *Rays of positive electricity and their applications to chemical analysis*. London, UK: Longmans Green.

Todd, J. F. J. (1991). Recommendations for nomenclature and symbolism for mass spectroscopy. *Pure and Applied Chemistry*, *63*, 1541–1566.

Wakankar, A. A., & Borchardt, R. T. (2006). Formulation considerations for proteins susceptible to asparagine deamidation and aspartate isomerization. *Journal of Pharmaceutical Sciences*, *95*(11), 2321–2336.

Wang, G., & Cole, R. B. (1994a). Disparity between solution-phase equilibria and charge state distributions in positive-ion electrospray mass spectrometry. *Organic Mass Spectrometry*, *29*, 419–427.

Wang, G., & Cole, R. B. (1994b). Effect of solution intic strength on analyte charge state distributions in positive and negative ion electrospray mass spectrometry. *Analytical Chemistry*, *66*(21), 3702–3708.

Wang, T., Chu, L., Li, W., Lawson, K., Apostol, I., & Eris, T. (2017). Application of a quantitative LC-MS multiattribute method for monitoring site-specific glycan heterogeneity on a monoclonal antibody containing two N-linked glycosylation sites. *Analytical Chemistry*, *89*(6), 3562–3567.

Whitehouse, C. M., Dreyer, R. N., Yamashita, M., & Fenn, J. B. (1985). Electrospray interface for liquid chromatographs and mass spectrometers. *Analytical Chemistry*, *57*(3), 675–679.

Wiley, W. C., & McLaren, I. H. (1955). Time-of-flight mass spectrometer with improved resolution. *The Review of Scientific Instruments*, *26*, 1150.

Wilm, M., & Mann, M. (1996). Analytical properties of the nanoelectrospray ion source. *Analytical Chemistry*, *68*(1), 1–8.

Wilson, S. R., & Wu, Y. (1993). Applications of electrospray ionization mass spectrometry to neutral organic molecules including fullerenes. *Journal of the American Society for Mass Spectrometry*, *4*(7), 596–603.

Yamashita, M., & Fenn, J. B. (1984a). Electrospray ion source. Another variation on the free-jet theme. *Journal of Physical Chemistry*, *88*(20), 4451–4459.

Yamashita, M., & Fenn, J. B. (1984b). Negative ion production with the electrospray ion source. *Journal of Physical Chemistry*, *88*(20), 4671–4675.

Young, P. M., & Wheat, T. E. (1990). Alternative mobile phases for enhanced chromatographic selectivity and increased sensitivity in peptide separations. *Peptide Research*, *3*(6), 287–292.

Zubarev, R., & Makarov, A. (2013). Orbitrap is the newest addition to the family of high-resolution mass spectrometry analyzers. With its revolutionarily new, miniature design, Orbitrap combines high speed with excellent quantification properties, ranking favorably in many analytical applications. *Analytical Chemistry*, *85*, 5288–5296.

CHAPTER 5

Separation techniques for characterization and quantification of monoclonal antibody purity

Lowell J. Brady
Drug Product Analytical Development, Drug Product Services, Lonza AG, Basel, Switzerland

5.1 Introduction and overview

Monoclonal antibodies are multisubunit proteins of approximately 150 kDa and comprise two light chains (LCs) and two heavy chains (HCs), each of which is involved in antigen binding (Chiu, Goulet, Teplyakov, & Gilliland, 2019). In biology, antibodies play a key role in immunity through recognition of foreign antigens, allowing organisms with antibody-based immunity to fight off infection and disease. The ability of antibodies to recognize specific protein structures with very high specificity combined with potentially long half-lives has led to the use of recombinant antibodies for the treatment of a diverse range of human diseases.

The first reported generation of an engineered monoclonal antibody–producing cell-line was described by Kohler and Milstein (1975). Combined with the rise of recombinant DNA technology and other advancements in biotechnology, the use of monoclonal antibody technology led to the first licensed drug of this class, muromonab-CD3 (Thistlethwaite et al., 1987), in 1986. Further developments, including antibody humanization, Chinese hamster ovary (CHO) cell production, and the development of purification (Shukla, Hubbard, Tressel, Guhan, & Low, 2007) and analytical platforms (Flatman, Alam, Gerard, & Mussa, 2007), have led to the rise of monoclonal antibodies as the dominant class of protein therapeutics and enabled new approaches in the treatment of a wide range of human illnesses. Monoclonal antibody drugs are used across a range of therapeutic indications, from cancer (Tang, Yu, & Hubbard-Lucey, 2018) to chronic indications such as rheumatoid arthritis (Rein & Mueller, 2017).

Monoclonal Antibodies
DOI: https://doi.org/10.1016/B978-0-12-822318-5.00008-9
Copyright © 2021 Elsevier Inc.
All rights reserved.

In concert with the rise of monoclonal antibodies as therapeutics, advances in microelectronics (Brock & Moore, 2006), precision manufacturing, improved quality and homogeneity of column matrix components, the availability of advanced optics and profound developments in the capability of computer software have enabled advances in separation sciences that have resulted in the development of high-resolution analytical techniques, from liquid chromatography to mass spectrometry (MS). The wide availability and standardization of equipment for performing analytics of proteins has resulted in a concomitant standardization of separations science methodology and normalization of techniques that have been applied for the analytical testing of monoclonal antibodies. In addition, due to the modular subdomain structure of these proteins, relatively similar molecular properties from antibody product to antibody product have allowed the development and use of platform approaches for cell-line development; purification and analytics have allowed development timelines to be reduced; and platforms have become a standard feature of antibody-drug development. By leveraging the similarities among members of this class of biologics with regard to physicochemical properties, stability, and protein structure, a set of standard analytical techniques, mostly based on separations science, has emerged for the determination and quantification of antibody product purity. The analytical data produced by these techniques are a routine aspect of the testing of monoclonal antibodies during both development and manufacturing phases and are an essential element of the chemistry, manufacturing, and control sections of antibody regulatory filings. For most antibody products, although the specific technique used to quantify and control product purity during routine measurements may vary from molecule to molecule or among regulatory submissions, there is high similarity with regard to the attributes and tests that are monitored and applied for most products (Ruesch et al., 2020). For nearly all monoclonal antibodies, techniques employing capillary electrophoresis (CE) and high-performance liquid chromatography (HPLC) or ultra-high pressure liquid chromatography (UHPLC or UPLC) are among the routine analytical methods that are commonly used for the quantification and characterization of antibody purity. For CE-based separation methods the use of chip-based separation systems has additionally enabled very high-throughput and robust separations that can be used for testing from process development to product release, and continued development with regard to instrument miniaturization and versatility is anticipated (Morbioli, Mazzu-Nascimento, Aquino, Cervantes, & Carrilho, 2016).

These techniques are most frequently used to separate and quantify product size and charge variants. With orthogonal testing the peaks separated by these methods can be used to characterize and quantify specific product quality attributes, including the individual amino acid changes (including product fragmentation) that may occur during the cell culture, purification, bulk processing, and primary packaging operations that comprise monoclonal antibody manufacturing, in addition to changes that may occur over time in drug substance or drug product stability studies.

The use of separation-based analytical techniques for the quantification of antibody purity and product-related variants is particularly useful because these methods may be used for both in-depth characterization as part of structure-function studies, and routine release and stability analysis in quality control (QC) environments. In many cases, analytical separations can be used either directly or with adaptation to both quantify and isolate product variants for further characterization using orthogonal analytical methods, including biological activity assays. The ability to directly isolate and characterize variants observed by separation techniques has helped to enable the development of critical quality attribute (CQA) classification systems (Alt et al., 2016; Goetze, Schenauer, & Flynn, 2010) across the biotechnology industry. These attribute assessment and classification systems seek to define in a systematic manner the degree and importance of the impact of each attribute and species observed as part of product quality and characterization testing and to utilize the relative criticality determination resulting from these studies in a quality-by-design framework (Finkler & Krummen, 2016; Kepert et al., 2016).

The versatility of separation-based methods for the determination of product heterogeneity is essential to the value of these methods as part of the control strategy and has led to refinement and further standardization of these techniques. This iterative process has led to a relatively standard suite of separation methods employed as part of antibody testing industry wide, resulting in a high degree of consistency with regard to how these methods are employed. Furthermore, as methodology and application of separation-based purity tests to monoclonal antibodies have become more standardized and a deep understanding of the product variants typically resolved and quantified by these methods has become available, the classification of attribute criticality and the impact of various impurities and product-related species quantified by these methods have also become harmonized across the biotechnology industry. Examples of this harmonization are the wide agreement on the importance of antibody aggregates (as discussed in Section 5.2) and relatively low impact of antibody

C-terminal HC lysine charge variants for intravenously administered products (Cai, Pan, & Flynn, 2011).

For most antibody molecules the standard separation-based methods that are used as part of product quality testing are size-exclusion chromatography (SEC); capillary electrophoresis sodium dodecyl sulfate (CE-SDS), under both reducing and nonreducing conditions; and charge variant separation, using one of three dominant techniques: cation-exchange chromatography (CEX), capillary zone electrophoresis (CZE), or isoelectric focusing (IEF). In addition, other analytical separation methods that are used during characterization studies or in support of process development include the following modes of separation: hydrophobic interaction chromatography (HIC), reversed-phase (ultra) HPLC (rp-(U) HPLC), anion-exchange chromatography (AEX), and boronate affinity chromatography (BAC). These techniques are less commonly used as part of the control system for monoclonal antibodies but may be used to answer specific analytical questions during product characterization or as tools for in-process testing support. In some cases, these methods may be used on a routine basis, for example, for product release and stability testing, when a monoclonal antibody has a less typically observed product quality attribute that must be routinely monitored.

Most of the techniques listed earlier are performed on antibodies without prior fragmentation or reduction during sample preparation and are intended to separate and quantify quality attributes of the intact molecule. Of all separation techniques typically included on the control system for routine quality monitoring of antibody purity, only reduced CE-SDS is based on analysis at the subunit level. Separation methods using antibody reduction or proteolytic cleavage are commonly used during characterization studies and may be employed for identity testing in routine operations. An in-depth discussion of separation techniques that can be applied to fragmented or digested antibody fragments is beyond the scope of this chapter, but the use of these methods is referenced as appropriate in the context of orthogonal characterization of the primary separations techniques that are discussed.

At many large organizations routinely utilizing these techniques for the testing and control of many different antibody products, standard ("platform") analytical methods have become an essential strategic element used to reduce development timelines and enable streamlined approaches to apply these methods in both development and for QC testing under good manufacturing practice requirements (Lacher, Roberts et al., 2010;

Michels et al., 2015; Moritz et al., 2015; Smith, Zhang, Adams, DiPaolo, & Dally, 2017). This approach is possible because of the relatively small degree of variation in antibody behavior as assessed by these standardized separation-based purity methods. Typically, an abbreviated assessment of the suitability or "fit to platform" for each molecule is performed to determine whether the particular monoclonal antibody in question has performance characteristics when assessed by the method under evaluation that demonstrate acceptable performance within predefined parameters. Successful assessment of suitability typically allows the further use of the method for analytical testing at early product development phases, including within QC, without performing product-specific development, or complete validation activities.

This chapter will provide an overview of the main separation methods used for monoclonal antibody testing and their application and use as part of the control system for antibody product characterization and quality control testing, including the definition of attribute criticality through isolation and characterization of the species resolved by these techniques. Considerations and use of less commonly employed separation techniques for specialized applications and less commonly encountered product variants will also be presented. Where appropriate, the context about the operation of these methods as part of a platform framework for monoclonal antibodies will also be discussed.

5.2 Size-exclusion chromatography

Size-exclusion chromatography (also called "gel filtration chromatography" as well as a variety of other terms) is a separation method based on the resolution of analytes through differential permeation of a porous chromatography matrix. Early separations of biomolecules were attained using starch as a separation matrix (Lindqvist & Storgårds, 1955). Most SEC columns in use for protein separations today employ either silica or hybrid organic/inorganic particles as further discussed below. Detection of the analyte is typically performed by ultraviolet (UV) light absorbance detection at either 280 nm, based on the absorption of aromatic amino acids, or in the range of 214 nm, based on the absorbance of the peptide bond. The choice of detection mode is dependent on the level of the analyte applied to the column and the potential for interfering excipients in the formulation or mobile phase to absorb and prevent accurate analyte quantitation. For antibodies, due to the sufficient number of aromatic residues and relatively large protein size, either of the UV detection options is possible. Alternative approaches that are often used to increase sensitivity, such as fluorescence detection, are usually not required but may be selected based on laboratory preference or to improve baseline stability or increase signal-to-noise. Fluorescence detection may also be used in special situations when products are prepared at low concentration, for example, during testing of samples for clinical administration studies. Irrespective of the detection approach, analyte peaks are usually integrated through comparison to a blank injection or reference chromatogram (to exclude inclusion of buffer-related peaks) and reported based on relative percent purity as compared to the main peak.

Most SEC chromatography columns used for HPLC-based analytical separations of antibodies are based on a rigid silica matrix. Although HPLC SEC columns composed of organic gels are also available, these kinds of columns are not typically used for analytical separations of monoclonal antibodies due to lower pressure tolerances and flow rates required for separations, which do not lend these types of columns to high-throughput separations needed for most antibody testing. Silica-based columns used for antibody separations are typically derivatized with diol functional groups to shield interaction of the analyte with silica silanol groups to limit mixed-mode interactions (Roumeliotis & Unger, 1981). The pore size of the chromatography matrix is selected based on the size of the analytes to be separated. Analytes with a larger hydrodynamic radius are less likely to penetrate the pore of the stationary phase, resulting in faster elution. Conversely, analytes with relatively smaller hydrodynamic radii enter the pores of the stationary phase more frequently, resulting in later migration (Hong, Koza, & Bouvier, 2012).

The development of ultra-high pressure chromatographic separations for various types of separation modes has also led to significant development in the availability of new types of SEC columns, employing alternative particles to the diol-modified silica-based HPLC columns as discussed earlier. UPLC separations require column packing materials that can withstand higher pressures, which has led to the development of column matrices such as the bridged hybrid ethyl particle, which can be surface modified with diol groups and offers similar or better performance relative to standard silica columns (Brusotti et al., 2018).

For antibodies, since hydrodynamic radius is generally directly proportional to molecular weight, SEC is used to separate and quantify size variants, especially those due to molecule dimerization or aggregation. These species are usually referred to as aggregates or high-molecular weight (HMW) species (Joubert, Luo, Nashed-Samuel, Wypych, & Narhi, 2011). The use of SEC for the quantification of molecule aggregates typically offers higher performance for monoclonal antibodies than for molecule fragments, as baseline resolution of molecule fragment species may not always be achieved and is dependent on the nature of the particular species in question (Li et al., 2017). Employing UPLC columns and instrumentation for SEC separations of antibodies typically offers higher performance relative to HPLC separations due to the combination of smaller column packing material particle sizes and smaller instrument dead volumes and may result in significant separation improvements (Yang et al., 2015).

SEC is generally applied to all antibody molecules during both development and QC phases of the product development lifecycle, from cell-culture optimization (Gomez et al., 2012), through purification development (Cromwell, Hilario, & Jacobson, 2006; Doss et al., 2019; Jin et al., 2019) and for release of clinical and commercial batches. For antibody products, HMW species are considered a CQA primarily due to concerns about immunogenicity (Bessa, Boeckle, & Beck, 2015; Filipe et al., 2012). HMW species for antibodies are typically measured by SEC at release and on stability, with product-specific limits established and revised as appropriate once a sufficient number of batches are produced. Isolation and characterization of HMW species are an important part of the product analytical lifecycle and the results of characterization tests performed on these species may reveal a variety of underlying reasons for aggregates to form, including both noncovalently (Bickel et al., 2016) and covalently (Brych et al., 2010; Xu et al., 2017) mediated forms. As the HMW peak or peaks resolved by SEC for antibodies may contain aggregate species due to more than one underlying cause, isolation and characterization of these species from product to product are typically performed, especially in later stages of development, and may reveal both covalent and noncovalent aggregate species within the same antibody batch or product (Kükrer, Filipe, & van Duijn, 2010) and even within the same peak as resolved by SEC.

Optimization of SEC methods typically includes selection of the column; determination of the composition, pH, and ionic strength of the mobile phase; the flow rate of the separation; load of the analyte; and the choice of using a standard HPLC- or UPLC-based separation approach. Since most separations are conducted in aqueous buffers, a low concentration of a bacteriostatic additive such as sodium azide is typically added to the mobile phase and column storage solution. Increased use of UPLC separations for SEC, including in QC laboratories, has become much more common for monoclonal antibodies with the advent of small-particle [sub-3 μm (Goyon, Beck et al., 2017)] separation columns and liquid chromatography systems capable of operating under the high pressure needed to operate these columns. Given that SEC separation is achieved based on permeation of the analyte through the porous column matrix, as opposed to binding of the analyte to a column functional group with subsequent elution, band broadening is an important consideration. UPLC may offer advantages in terms of reduced separation time, increased resolution, or both, as compared to HPLC (Mou, Yang, Li, Ambrogelly, & Pollard, 2014), depending on the specific comparison made and the specific settings selected for the separation (Fekete, Veuthey, & Guillarme, 2017). Irrespective of whether HPLC or UPLC instrumentation is used, limiting system dead volume and optimizing the analyte injection volume are important considerations to reduce band broadening, which is one of the main performance-related challenges that can be observed with SEC separations (Ghrist, Stadalius, & Snyder, 1987; Goyon, Guillarme, & Fekete, 2017). For this reason, injection volume size is usually limited to increase analytical performance and product variant resolution, and instruments are typically optimized by limiting the diameter and length of tubing, avoiding unnecessary valves, and placing the detector flow cell close to the outlet of the chromatography column. As discussed further earlier, the capability of UPLC systems to make μL or sub-μL injections may also offer the option of injecting product at the intended drug substance or drug product concentration without detector saturation, which may be important for understanding concentration-dependent aggregation levels. The ability to inject very small volumes also provides an advantage of UPLC-based methods relative to HPLC method given the ability to analyze very small sample quantities.

Most SEC separations for antibodies are conducted in a pH range of approximately 6.0–7.0, with consideration of the product pI taken into account for selection and optimization of the final pH condition. Typically, phosphate-based buffers, with or without sodium chloride to reduce ionic interactions, are employed (Goyon, Beck, Veuthey, Guillarme, & Fekete, 2017). The addition of mobile phase additives such as arginine (Ejima, Yumioka, Arakawa, & Tsumoto, 2005; Yumioka, Sato, Tomizawa, Yamasaki, & Ejima, 2010), sodium perchlorate (Wang, Levi, Del Grosso, McCormick, & Bhattacharyya, 2017), cyclodextrin

(Javeri et al., 2018), or organic additives such as isopropanol may be considered, particularly when interaction of the product with the column matrix is observed (Arakawa, Ejima, Li, & Philo., 2010). However, addition of modifiers such as arginine that may impact the observed level of aggregation due to chaotropic effects should be carefully evaluated to ensure method accuracy (Tsumoto, Ejima, Kita, & Arakawa, 2005). Assessment of the impact of added buffer components on the accuracy of HMW species can be difficult to assess due to lack of ready orthogonal techniques to evaluate product aggregates in the native (nondenatured) state. Alternative techniques that may be used for this purpose for monoclonal antibodies include asymmetric field flow fractionation (AF4) (Boll, Josse, & Heubach, 2018) and analytical ultracentrifugation (Gabrielson et al., 2007) but these powerful techniques are typically employed only for characterization purposes and are not typically used as part of QC release and stability testing. In addition, due to the inherent differences in the quantification of antibody size variants by these techniques, it is usually only possible to make general statements about the degree of agreement with regard to aggregation level for a product, particularly when aggregation level is relatively low, which is usually the case for monoclonal antibodies. However, if sufficient time and resources are available, evaluation of these techniques and inclusion of the data from them as part of the effort to understand antibody aggregates for a specific product can provide important context for interpretation of routinely generated SEC chromatography data. For both AF4 and SEC separations, coupling of the chromatography column to a multiangle light scattering (MALS) detector can be used to generate molecular mass data for the separated peaks (Esfandiary et al., 2013). The hyphenation of SEC and MALS can be particularly useful when column matrix interactions are suspected to evaluate the correspondence of the MW determined from MALS calculations (an absolute method) relative to those estimated based on calibration of the SEC column with protein molecular weight markers (a relative method). In addition to characterization of aggregate and fragment products resolved by SEC or AF4 by light scattering, the collection of aggregate peaks and characterization by a range of other assays, from MS to bioassay testing, may also be performed and commonly a part of the last stage product characterization in support of regulatory submissions.

One alternative form of SEC that can be used to characterize size variants is to operate the column in the presence of a chaotropic salt such as guanidine hydrochloride. Denaturing SEC (Van Buren, Rehder, Gadgil, Matsumura, & Jacob, 2009) can be used to separate antibody variants under either nonreducing or reducing conditions, as the presence of the denaturant maintains the solubility of reduced antibody chains within the chromatography column after separation from the sample preparation buffer, which usually contains a mixture of reducing agents such as diothiothreitol or β-mercaptoethanol in a denaturing buffer. Using this approach, species may be resolved and collected for further analysis by further orthogonal techniques such as MS or CE-SDS (Michels, Brady, Guo, & Balland, 2007).

In addition to chromatography hardware considerations, sample handling, especially dilution prior to injection, is an additional important aspect to consider as part of method optimization for monoclonal antibodies. For some products the degree of aggregates observed as a result of the separation may vary based on product concentration, typically with higher levels of aggregates observed at higher antibody concentrations (Liu, Nguyen, Andya, & Shire, 2005; Nicoud et al., 2016). This observation can often lead to challenges during method development and analytical testing, as methods analyzed after predilution of the product may result in both a lower level of quantification of product HMWs and the potential for variation in the observed level of HMW that are quantified. This may be particularly problematic if the equilibrium processes for formation of HMW species after product dilution operate at timescales greater than those observed as part of analytical testing. Depending on whether an intravenous or subcutaneous mode of administration is employed for dosing, the relationship between the sample handling procedure used for analytical testing and potential impact on quantified HMW species may be more or less relevant. A thorough understanding of product aggregation behavior as a function of concentration and evaluation of potential aggregate reversibility (Joubert et al., 2011) is an important consideration during method selection and product development. These questions may be addressed in part by the use of clinical administration studies, in vivo models (Fathallah et al., 2015; Kijanka et al., 2016), or assessment of product behavior in actual or simulated human serum (Luo et al., 2020) experiments.

5.3 Charge-based separation methods

Separation of protein variants by exploiting variations in analyte charge is a technique used both during large-scale purification and at analytical scale for purity determination of the isolated product. For monoclonal antibodies, resolution and quantification of antibody-charged isoforms are an important part of the product control system for every monoclonal antibody molecule during all stages of product development, from cell-culture development (Tang, Miao, & Zhang, 2018; Trappe et al., 2018) to commercial manufacturing (Tebbey, Varga, Naill, Clewell, & Venema, 2015). Although various techniques based on chromatographic or electrophoretic separation methodologies exist for the separation of antibody-charged isoforms, these techniques share similar methodological outputs and typically result in the resolution of the sample into one major peak and several minor isoforms bearing more acidic or more basic relative charge. The charged-isoform pattern obtained across the product development lifecycle is an important tool for monitoring the consistency of antibody production as well as changes over time on drug substance and drug product stability (Gandhi, Ren, & Xiao, 2012; Lee et al., 2018) and is an important part of assessing the impact of process changes on product quality.

Although the separation techniques usually employed cannot determine the nature or biological relevance of the separated variants, characterization of the product profile resulting from these methods with orthogonal characterization tests, particularly MS-based methods, allows the composition of charge variant species, often at the amino acid level, resolved by these methods to be determined (Perkins, Theiler, & Lunte, 2000). Antibody charge variants are not presumed to be obligatory (mandatory) CQAs but rather are considered product-related species and are not typically considered impurities, unless determined to be so as the outcome of characterization studies. Once variants are purified, characterized, and tested by a panel of orthogonal techniques, the relative criticality of the different charged isoforms can be determined, typically by applying a scoring system to determine if the modification or variant is a CQA. It is important to note that identification of the underlying molecular cause of a particular charged variant is usually not sufficient to determine attribute criticality and is but one part (Li et al., 2016) of the attribute criticality assessment process, which usually includes biological testing of isolated species, if feasible.

Commonly detected acidic variants of monoclonal antibodies include deamidation of asparagine (Vlasak et al., 2009) residues to either aspartic or isoaspartic acid (Sydow et al., 2014), or both; the presence of charged glycans (as either *N*-acetyl neuraminic or *N*-glycolyl neuraminic acid forms) (Sheeley, Merrill, & Taylor, 1997), depending on the expression system and source of raw materials (Diaz et al., 2009); variants due to disulfide reduction (Miao, Xie, & Zou, 2017); and sequence variants resulting in net negative charge addition, including product fragments. Typically encountered basic variants include the presence of C-terminal HC lysine variants due to incomplete enzymatic processing during cell-culture production; C-terminal amidation, typically of the penultimate proline residue of the HC (Kaschak et al., 2011); stable succinimide formation; and the presence of sequence variants (including N-terminal glutamine cyclization variants) resulting in net positive charge addition (Dada, Jaya, Valliere-Douglass, & Salas-Solano, 2015). The presence of charge isoforms due to sequence variants (Ren et al., 2011; Scott, Rogers, Balland, & Brady, 2014) may result in either acidic or basic forms as the result of product fragmentation, amino acid replacement during cell culture, or the presence of multiple product sequences due to nonclonal cell lines. Other variants observed that may also be resolved by charge-based techniques include modifications due to oxidation and glycation (Quan et al., 2008), including advanced glycation end products (Liu, Madren, Feng, & Sosic, 2020).

As mentioned earlier, three main types of charge-based separation methods are typically employed for the separation of antibody charge variants: cation-exchange HPLC (CEX-HPLC) (Lingg et al., 2014), isoelectric focusing (IEF), and CZE (Kahle & Wätzig, 2018) A comparison of these techniques is shown in Fig. 5.1. Isoelectric focusing may be performed using electrophoretic systems where the focused analytes are separated in the capillary and then mobilized and detected as they pass a detector window (Salas-Solano, Babu, & Park, 2011) or by focusing in a capillary that has a detector window covering the entire separation region, referred to as "imaged" isoelectric focusing (Wu, Yu, Wang, & Wang, 2018). Imaged isoelectric focusing (icIEF) has the added advantage that the process of focusing can be observed using analysis software, which can enable method development and optimization. With the advancement in column manufacturing and the capability of chromatographic systems to operate at ultra-high pressure, UPLC separations have become more common and may take the place of HPLC-based CEX separations. Selection among CEX, CZE, or IEF techniques is often based on molecule performance in the assay type or adherence to a particular technology platform within the organization developing the methodology. One consideration for the selection of separation techniques for resolution of charged variants is the differential principles for separation in each mode: chromatographic

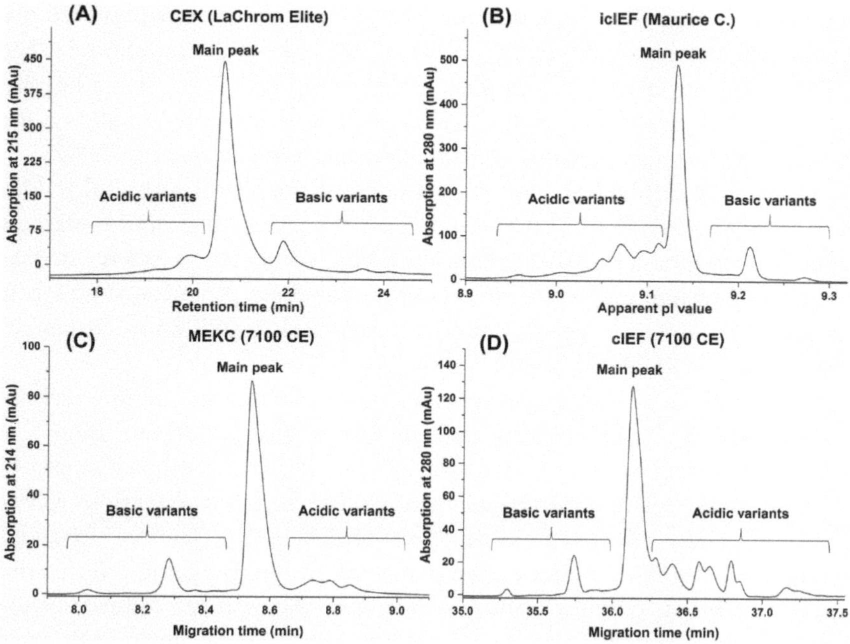

Figure 5.1 The figure above shows a comparison of chromatographic and capillary based charge separation techniques applied to a monoclonal antibody. The similarities and differences among the profiles obtained demonstrate the selectivity that can be obtained by applying different charge-based separation techniques to the same monoclonal antibody sample.

separations typically separate only based on surface charge, whereas electrophoretic separations resolve charged isoforms based on total net charge. This difference in separation principle may result in different profiles for the same antibody when separated using different techniques and can be exploited to further understand the nature of charge variants when multiple modes of charge variant separation are compared for the same molecule (Kahle, Zagst, Wiesner, & Wätzig, 2019).

Although AEX may often be used to resolve charged variants of classes of protein therapeutics, particularly those with significant negative charge due to sialylated *N*- and *O*-glycosylation, sulfation, or phosphorylation, these charged posttranslational modifications are not routinely observed on monoclonal antibodies at significant levels. Although monoclonal antibodies contain a consensus N-linked glycosylation site in the Fc domain, the glycans that occupy this site do not typically contain significant degrees of N-linked sialylation; when sialylation is observed on the Fc domain, it is typically at a low level (<10%) and the species observed are typically mono-sialylated (Brady et al., 2015). Even for antibodies that contain an *N*-glycosylation site outside the Fc domain that is more heavily sialylated, such as cetuximab, CEX has still been successfully employed for the separation of glycosylation variants (Váradi, Jakes, & Bones, 2020).

Since most antibodies possess pI values in the basic range (pI values from approximately 6 to 9) (Goyon et al., 2017; Lingg, Tan, Hintersteiner, Bardor, & Jungbauer, 2013), anion-exchange separation is rarely used as the primary charge quantification technique for these products, although it has been used to solve specific antibody product characterization challenges (Teshima et al., 2011; Zhou et al., 2020).

For all three of the major types of charged-based separation techniques, detection of the antibody isoforms is typically performed using UV absorbance detection in the range of 214 or 280 nm and product purity determined by relative purity calculation. The selected detection wavelength is usually selected as a function of the analyte load or to avoid background absorbance due to the presence of absorbing moieties in the separation buffer or electrophoresis solution. Alternatively, fluorescence detection may be employed by taking advantage of the intrinsic fluorescence of antibody tryptophan residues, offering advantages with regard to signal to noise and baseline stability. As for SEC above, integration is usually performed by comparison to a blank, buffer matrix, or reference injection. By calculating the fractional area of each peak or peak region as a percent of the entire region, the contribution of each particular species resolved by the method to the molecule purity profile can be determined.

5.3.1 Cation-exchange chromatography

CEX methods have been employed for both characterization and QC purposes for decades and were used for some of the earliest monoclonal antibodies that were commercialized (Harris et al., 2001). CEX-HPLC separations can exist as either weak or strong forms using cationic exchangers as functional groups of the chromatography media. For analytical separations of antibody variants the selection of either a weak or strong cation exchanger is based on the behavior of the molecule and the particular mobile phase system selected, with either option providing the potential for fine-tuning the separation conditions to enable resolution of antibody variants differing in only a single charged residue. In weak CEX a neutral stationary media is functionalized with a weak cation exchanger such as carboxylic acid, whereas in strong cation exchange a stronger exchanger, such as sulfonate, is used. In analytical CEX chromatography, elution and separation of the product variants may be achieved using either salt or pH-based gradients. For elution using a salt gradient the product is bound to the column in a weak buffer and is selectively eluted using a salt gradient of a stronger buffer (Fekete, Beck, Fekete, & Guillarme, 2015a). Alternatively, binding may also be achieved by binding the antibody at a relatively low pH (below the pI value of the most acidic variant) and elution of analytes achieved by using a pH gradient (Fekete, Beck, Fekete, & Guillarme, 2015b), where each charged isoform elutes once the gradient pH and pI are equal. In both separation modes, secondary interactions can interfere with the separation and result in poor chromatographic performance. Irrespective of the particular approach selected for elution of the antibody, optimization of the elution gradient and mobile phase composition allows the separation and quantification of acidic, neutral, and basic charged components.

Development of a product-specific CEX method is typically performed through column selection; buffer composition, mode of elution, and gradient optimization; and choice of an HPLC or UPLC chromatographic system. Selection of the initial conditions may be based on previous experience with the product, available analytical hardware, and insight into molecule behavior based on protein bioinformatics data and during purification development. Design of experiment (DOE) techniques can be useful for both method development and robustness determination (Joshi, Kumar, & Rathore, 2017).

The two most commonly encountered challenges observed during CEX method development are proper optimization of the elution gradient and secondary interaction of the product with the column matrix. In the first case, optimization of the separation conditions, including buffer selection (Farsang et al., 2019) and flow rate, may be used to improve separation and achieve a robust and reproducible separation. Antibodies with extremes of pI may present particular difficulty in terms of obtaining sufficient resolution to obtain a suitable and reproducible peak profile. For pH gradient systems, adjustment of the ionic strength of the separation buffers may be a helpful modification to achieve a useful separation (Zhang, Patapoff, Farnan, & Zhang, 2013), particularly in these cases. In contrast, secondary interactions with the column matrix, which may result in excessive peak tailing or poor resolution, or both, may be more difficult to resolve. If not resolved by adjustment of the buffer system, evaluation of an alternative column (such as changing from strong to weak cation exchange in pH separations) may be useful. Where suitable peak shape is achieved but resolution is not deemed to be sufficient, UPLC separation may offer improved resolution. When evaluation of various chromatographic conditions has not resulted in sufficient product resolution, consideration of an alternative, electrophoretic-based separation technique may be warranted as discussed below. Alternatively, when successful HPLC separation has been achieved, a UPLC-based separation using an equivalent column matrix may allow for a faster separation, allowing savings with regard to equipment utilization, buffer consumption, and time. For both UPLC and HPLC separations, DOE approaches can be used for method development, method refinement, and method robustness confirmation (Kochling, Wu, Hua, Guan, & Castaneda-Merced, 2016).

Characterization of CEX-based antibody separations is an important activity that is routinely performed to understand the importance of quality attributes of the product and is essential for the determination of attribute criticality. CEX separations developed for analytical purposes can often be modified to run in a semipreparative mode while maintaining a similar separation capability to the analogous method operated at analytical scale. By purifying fractions from the separation, charge variants may be analyzed by orthogonal techniques, including biological assays (Fekete et al., 2015a). These experiments are crucial to understanding the criticality of resolved isoforms of monoclonal antibodies and can provide a rationale for which peaks resolved in the analytical CEX separation should be monitored routinely at product release and over stability, as well as the stringency of the limits applied to control the variation of these species. This can be particularly important at later stages of development, where monitoring of charge variants as regions (e.g., acidic and basic regions as a whole) may be insufficient and deeper understanding of individual peaks is usually required.

5.3.2 Capillary isoelectric focusing

Electrophoretic-based charge separation methods may be used in place of chromatographic ones, or in supplement, depending on the particular testing strategy applied. In contrast to CEX, IEF and CZE separations are based on differences in net molecule charge. Gel-based isoelectric focusing has been used as both a characterization and QC method for complex protein therapeutics and was applied for some of the first glycoproteins commercialized by the biotechnology industry (Battersby, Ohlrogge, Feigler, & Holloway, 1992). In tandem with the development of advanced HPLC and UPLC systems over the past several decades, significant improvements in the manufacturing and performance of CE systems have enabled the standardization of these techniques for the analysis of macromolecules, including monoclonal antibodies.

For capillary isoelectric focusing analysis of antibodies (Smadja, Tran, & Tavernaa, 2017), separation of the antibody charge variants is achieved through application of high voltage to a solution containing a mixture of analyte and ampholyte molecules. Upon application of voltage, the ampholyte mixture, which is a complex solution of different small molecules, establishes a pH gradient across the capillary. Antibody variants migrate to their isoelectric point, which is the location in the gradient where they have no net charge, resulting in the separation of molecule variants. In so-called cIEF the focused gradient is mobilized past the detector through aid of a chemical or hydrodynamic mobilizer. Alternatively, in imaged capillary IEF (icIEF) the separation is performed in a UV transparent capillary, and the entire focused region is detected using a capillary imaging system. In both types of capillary IEF separations, detection is usually performed by UV absorbance at A280 nm. Systems employing fluorescence detection may be also be used, which allow detection by excitation and intrinsic fluorescence of molecule tryptophan moieties. Ampholytes are selected for their lack of absorbance in the region of 280 nm to allow for analyte detection. In contrast, A280 absorbing molecules of known pI are used, typically in a bracketing manner, as pI markers to visually define the pI gradient. These pI markers can then be used to calibrate the pI gradient established by the separation and to calculate the pI values where each analyte peak is observed. After separation and imaging, the profile of the antibody peaks can be integrated and the relative purity of each variant calculated.

Commonly encountered challenges with separations of monoclonal antibodies by IEF include poor focusing, poor peak shape, and poor signal to noise for minor product variants. One commonly encountered challenge is the tendency of analytes to have lower solubility at their isoelectric points, which may be observed through the presence of inconsistent peak shape or the presence of unstable signal in the region of the detected product peaks. When precipitation is suspected, inclusion of a chaotropic agent such as urea may be considered, although care must be taken to ensure that degraded urea solutions are not used (Mack & Darling, 2014), which may cause product carbamylation and result in inaccurate and variable results. As an alternative to urea the use of formamide has also been reported and successfully used for the analysis of monoclonal antibodies (Zhang, Voronov, Mussa, & Li, 2017). To optimize product signal-to-noise, antibody loading studies can be performed to demonstrate a sufficient product load, which is particularly challenging in capillary-based IEF separations where contributions from the ampholytes may contribute to a noisy baseline and limit the reproducible quantification of minor species. Optimization of the ampholyte mixture used to establish the pI gradient is one of the most important aspects of both cIEF and icIEF method development. Typically, a wide range ampholyte mixture is used during the initial stages of development to ensure that the pI span of the product peaks has been properly understood prior to assessing the potential to improve species resolution through selection of ampholyte mixtures with narrower pI ranges. Where commercial ampholyte mixtures lack the proper pI range to enable the desired resolution for the species of interest, mixtures of different ampholyte products with narrower ranges can be used to build a tailored solution for a specific product during method development. Optimization and assessment of mixtures can be significantly aided by statistical approaches for data analysis, including DOE (Kahle, Stein, & Wätzig, 2019). It is important to consider that variation in batch-to-batch quality and performance of pI markers and ampholytes should be carefully tested during method development to ensure that system suitability criteria are properly selected and that a robust method has been developed. Changes in the availability of a specific pI marker or ampholyte product may provide operational difficulties for these methods during routine use, as equivalent alternatives within the scope of particular method developed for a specific product may not be readily available. When optimization of the amphoyte gradient, inclusion of urea and evaluation of different focusing and separation voltage ranges have not resulted in an acceptable charged variant profile, alternative techniques such as CEX-HPLC or CZE may be required for successful charge variant separation for the product.

Characterization of antibody variants separated by isoelectric focusing methods usually requires indirect methods, as separation and isolation by capillary-based techniques is usually not possible. Typically, an orthogonal semipreparative chromatography method such as CEX may be employed to isolate fractions and characterize them by orthogonal techniques, including by analysis using the capillary-based isoelectric focusing technique used for product testing. Alternatively, preparative methods based on the same principles as the analytical IEF separation technique may be used, including the use of the Rotofor (Hosken, Li, Mullappally, Co, & Zhang, 2016) device or use of preparative IEF-based methodology (Meert, Brady, Guo, & Balland, 2010). Both of these techniques have been successfully used to characterize antibody variants where chromatographic separation methods were not successful or did not resolve the variants of interest with sufficient resolution, but are usually of lower throughput and more labor-intensive relative to chromatography-based isolation methods.

5.3.3 Capillary zone electrophoresis

In CZE, analytes are separated in a capillary by application of an electric field in the presence of a separation matrix (commonly referred to as the background electrolyte). For monoclonal antibodies, separation is achieved based on the principle that different charged isoforms possess different charge-to-hydrodynamic-radii ratios. Under application of an electric field, the product variants migrate at different velocities within the capillary based on the charge-to-hydrodynamic-radius ratio, allowing differential detection of product variants bearing different charged species. Detection is usually performed by UV absorbance of the peptide backbone at 214 nm or a similar wavelength.

CZE separations may be optimized through the selection of the capillary and choice of capillary coating; pH of the background electrolyte; inclusion of charged modifier components such as sodium chloride to modify the velocity of the separation; inclusion of detergents or organic solvents to control hydrophobic interactions; and optimization of the composition of the background electrolyte (Shi, Li, Qiao, & Lin, 2012). DOE techniques have been effectively used for the optimization of CZE separations (Moritz et al., 2017). CZE separations have benefited from the same advances in instrument hardware that enabled increasingly robust capillary IEF and CE-SDS separations to be routinely employed in the biopharmaceutical industry over the past several decades. As for cIEF and icIEF, it is important to consider the robustness of developed methods and to evaluate the variation that may be introduced in a specific assay due to the source of the different chemicals used. If possible, the identification of multiple sources of specific key reagents should be included during method development to ensure robustness of the developed method and reduce risks of impacts due to changes in chemical availability or specific grades of chemicals available from suppliers.

CZE does not possess a routinely used semipreparative analog, requiring the development of orthogonal separation approaches for isolation of charged isoforms for product characterization. Typically, the similar approaches as taken for characterization of isoelectric focusing methods are taken, with re-injection analysis of the isolated variants onto the analytical CZE method to define desired fraction purity and verify the variants of interest have been successfully isolated.

Both CEX and CZE may be adapted so that direct interfacing with MS detectors is possible (Jooß, Hühner, & Kiessig, 2017; Shi, Xiao, Dillon, Ricci, & Bondarenko, 2020). Interfacing charge-based methods to MS usually requires a sacrifice with regard to some aspects of robustness and resolution obtained as compared to routinely implemented versions of these methods (e.g., as compared to QC release and stability testing). However, the opportunity to interface a charge-based separation directly with MS detection offers the ability to directly characterize charge variants without performing off-line isolation studies. It should be noted, however, that part of the standard approach for characterization of charge variants and determination of criticality as discussed earlier requires assessment in a panel of assays, usually including biological functional assays, which is not enabled by direct MS detection characterization approaches. If characterization studies requiring orthogonal assessment are required, isolation of variants may still be required even when direct interfacing of charge-based methods to MS is possible.

5.4 Capillary electrophoresis sodium dodecyl sulfate separation

SDS-polyacrylamide gel electrophoresis is a classic method for the separation of antibodies and other proteins for the quantification of product size variants under both reducing and nonreducing conditions. Although this method is still routinely employed in the biotechnology industry, especially in the very early stages of the product lifecycle during cell-line development, it is no longer routinely used as a product quality test for monoclonal antibodies during clinical development. Instead, this technique has been replaced by CE-SDS (Rustandi, Washabaugh, & Wang, 2008), which offers significant advantages in terms of sensitivity, quantitation, and reproducibility. As for the shift from the use of IEF gels for protein separation to capillary-based methods, significant improvements in the manufacturing and performance of CE systems has allowed polyacrylamide gel-based antibody separations to be replaced by much more robust, precise, and quantitative CE-SDS methods (Kahle, Maul, & Wätzig, 2018).

CE-SDS methods are routinely employed during product development and QC release and stability testing and are an important platform test for monoclonal antibodies to quantify size variants at both the intact and subunit (LC and HC) level (Sänger-van de Griend, 2019). In CE-SDS, analyte molecules are mixed with a sample buffer containing SDS and incubated at an elevated temperature, which is usually in the range of 70°C for monoclonal antibodies (Salas-Solano et al., 2006). A reducing agent, typically beta-mercaptoethanol, is optionally included if reduction of the antibody to its substituent LC and HC components is desired. Assessment of the suitability of sample preparation conditions is an important aspect of method development or platform method adaptation for monoclonal antibodies to avoid the presence of variable or spurious artifacts due to incomplete produce reduction (Zhang et al., 2020). This is of particular importance when the presence of nonreducible covalent linkages are suspected to be present in a product sample, as incompletely reduced antibody samples due to nonoptimized sample preparation conditions may interfere with accurate quantification of these species. For nonreduced separations, an alkylating reagent is included to avoid method-induced auto-reduction of the antibody chains (Fekete, Gassner, Rudaz, Schappler, & Guillarme, 2013), particularly for lambda LC-containing antibodies (Wagner et al., 2020). Here, similar care must be taken during method development to ensure that peaks observed in antibody samples under nonreduced conditions are not due to nonoptimized sample preparation conditions (Molina et al., 2020).

After sample preparation the sample is introduced to a capillary containing a sieving gel. The negatively charged SDS forms complexes with and unfolds the analyte molecules. The separation is obtained by applying voltage across the capillary, with subsequent migration of the charged SDS-analyte complexes past a detector window. Since larger antibody size variants complex with a larger number of SDS molecules, they have a higher net negative charge, resulting in faster migration through the separation gel. Detection is typically obtained by UV absorbance at 214 or a similar wavelength. Alternatively, the sample may be derivatized prior to separation with a fluorogenic dye and detection performed by excitation of the dye at the appropriate wavelength followed by detection in the emission region of the dye (Michels, Parker, & Salas-Solano, 2012). Detection using intrinsic fluorescence of tryptophan moieties (Couderc, Ong-Meang, & Poinsot, 2017) in the analyte may also be employed. Alternatively, the analyte-SDS moieties may be detected through complex formation with a fluorescence dye and detection enabled by microfluidic dilution of the dye-analyte-SDS complexes following separation to overcome background fluorescence of nonspecific dye-SDS complexes (Chen, Ha, & Rustandi, 2013). CE-SDS separations are compatible with microfluidic and chip-based techniques, and these high-throughput methods are routinely used in monoclonal antibody separations, including in QC laboratories.

For monoclonal antibodies, CE-SDS is an important assay for the determination of product size variants, both during product characterization and release testing and during monitoring on stability. Under reducing conditions, the integrity of the LC and HC may be monitored and the formation of product fragmentation variants detected. In addition, reduced CE-SDS separations may also be used to analyze nonglycosylated HC variants (Chen, Tang, Lee, & Flynn, 2008), whether due to production by cell-culture or enzymatic treatment, glycosylated HC variants (Valliere-Douglass et al., 2010), formation of nonenzymatic glycation variants (Kaschak, Boyd, & Yan, 2011), or sequence variants (Harris et al., 2019). In addition, monitoring nonreducible linkages such as thioether bond formation is also possible in reducing CE-SDS analyses.

Nonreduced CE-SDS separation most commonly detects assembly variants of monoclonal antibodies produced through partial formation of disulfide bonds (Lacher, Wang et al., 2010). These variants may or may not be detected by orthogonal assays such as SEC, where size variants are usually not detected in cases where the antibody structure of two LCs and two HCs is held together by noncovalent interactions. Commonly observed antibody species detected by nonreduced CE-SDS include free LC, heavy-light, and heavy-heavy-light arrangements. Free HC and heavy-heavy variants are less commonly observed, although they may be present (Moritz & Stracke, 2017). CE-SDS is commonly used as a complementary method to SEC for size variant quantification and characterization and can be used to further understand the composition of size variants observed and collected from SEC fraction collection in conjunction with other methods such as MS (Boyd et al., 2018). Analysis of antibodies through a combination of nondenaturing SEC-based methods in combination with CE-SDS techniques (which are inherently denaturing) provides information about antibody size variants at multiple levels, from fragmentation (reduced CE-SDS) to aggregation (SEC) as well as the potential to detect nonreducible species (reduced CE-SDS) and covalent aggregates (nonreduced CE-SDS).

Separation of monoclonal antibodies under nonreducing conditions requires careful consideration with regard to sample preparation conditions. When denaturation in SDS is performed at basic pH (e.g., 8.5), the incubation process may allow free cysteine residues that are buried inside the antibody structure (Zhang et al., 2012) to be exposed to the solvent and to participate in disulfide re-arrangement reactions at the same timescale as the sample incubation. Failure to block this process by using alkylation reagents such as iodoacetamide or *n*-ethyl maleimide during the sample incubation may result in the artifactual observation of product fragments that are not an authentic component of the sample. Given that CE-SDS performed under nonreducing conditions may one of the few assays to detect product size variants due to disulfide reduction during the cell-culture harvest and capture procedure (Trexler-Schmidt et al., 2010), it is critical to employ sample preparation techniques that can authentically detect antibody product size variants observed under nonreducing conditions (Esterman, Katiyar, & Krishnamurthy, 2016).

The criticality assessment of the variants detected by CE-SDS separations is usually performed by generation of the species in other assays (such as chromatographic methods that allow peak fractionation), followed by assessment using the CE-SDS method established for the molecule. This is due to the difficulty of purifying separated variants from CE and the interfering nature of SDS-protein complexes produced by the method with orthogonal testing. Understanding the nature, and in particular the biological relevance, of the fragments or covalent aggregates observed by CE-SDS is an important aspect of defining the criticality of species observed by this method (Wang, Liu, Cai, Huang, & Flynn, 2015; Yang et al., 2019).

5.5 Hydrophobicity-based separation methods

As described earlier, the most commonly employed techniques for separation of monoclonal antibody purity variants are SEC, one of three primary types of charge variant separation methods, and CE-SDS operated under both reduced and nonreduced conditions. The combination of these four types of assays is usually sufficient to control antibody size and charge variants during routine testing as part of the control system for both product release and stability, for both drug substance and drug product. These tests are commonly employed as the main antibody purity tests for routine analysis and typically included on the test specification in combination with compendial and biological activity test methods.

In addition to these commonly used separation techniques, several additional methods that rely on hydrophobicity differences within the sample should be mentioned that are also important tools for the assessment of antibody purity. The most commonly used hydrophobicity-based separation methods are hydrophobic interaction chromatography (HIC) and reversed phase (rp)HPLC or UPLC (Martínez-Ortega et al., 2018; Yan et al., 2007). These separation methods are typically employed to answer specific questions during characterization studies or to serve as analytical tools during particular aspects of product development and less frequently as part of the standard release and stability assay panel. The need for additional, orthogonal separation assays to those previously discussed may arise due to the observation of particular product-specific modifications, such as those detected through MS experiments or through product characterization studies, such as those mentioned in the previous sections. In these cases the use of these assays beyond their typical characterization purpose as part of the control system may be required. Alternatively, these assays may be regularly employed during specific phases of the antibody development cycle, such as during candidate selection, where hydrophobic interaction chromatography (HIC) is commonly used to evaluate hydrophobicity variants and may be used to rank development candidates (Bailly et al., 2020) in terms of potential difficulty with regard to fit to purification or formulation platforms.

HIC separations are performed by binding the antibody in a high-salt buffer, usually containing ammonium sulfate, to a hydrophobic column matrix, with elution in step or gradient mode to reduce the concentration of the salt. Interactions of antibody amino acids that are promoted in the presence of ammonium sulfate are weakened through a decrease in ammonium sulfate concentration and subsequently eluted. Column matrixes based on butyl, phenyl, or ethyl amide chemistry are often chosen, but optimization for each particular antibody and the desired separation to be achieved is usually required (Valliere-Douglass, Wallace, & Balland, 2008). HIC separation may be analogous to rpHPLC separation in selectivity in the sense that hydrophobicity variants due to chemical modifications or protein-folding variants can be resolved, though usually with less resolution, but offers an important advantage since fractions taken from HIC chromatography separations usually retain biological activity when mobile phase components are removed. This enables the testing of product variants, that may not be successfully separated by other techniques, using orthogonal methods, including bioassays. HIC separations have been used to resolve less commonly observed antibody variants such as *O*-fucosylation (Valliere-Douglass, Brady, et al., 2008), for quantification and isolation of succinimide (Cao et al., 2019) and isomerization variants of aspartic acid and to quantify oxidized forms of tryptophan (Boyd, Kaschak, & Yan, 2011).

Separation of monoclonal antibodies by rpHPLC offers similar selectivity, though usually higher resolution, relative to HIC, and is routinely performed as part of interfacing chromatographic separation with MS. Separations achieved by rpHPLC are typically capable of separating size, sequence, glycosylation, and oxidation (Yang, Wang, Liu, & Raghani, 2007) variants, in addition to chemical modifications such as deamidation or oxidation. In contrast to HIC, however, collection of peaks from rpHPLC and testing by other assays are not usually employed for product characterization, as samples are usually denatured and lack native protein structure, thus presenting challenges with regard to product solubility in nondenaturing solutions, and thus preventing meaningful testing by many orthogonal techniques. The characterization and criticality determination approach for antibody species resolved by rpHPLC are similar to that taken for CE-SDS, where species or peaks generated by other methods are analyzed by rpHPLC as part of a panel of testing to assess attribute importance.

Reversed-phase separations may be performed without prior sample treatment, or after reduction, or reduction and alkylation, of the cysteine amino acids of the antibody. Nearly all covalent modifications of monoclonal antibodies can be analyzed using these techniques and detected by interfacing the separation with MS detection, though care must be taken to select sample preparation and separation conditions that do not create artifacts due to the analysis. Frequently, rpHPLC-MS separation is performed in-line with UV detection, usually at 214 nm, to provide an orthogonal means of quantification for peaks that are chromatographically separated. This versatile separation technique may be used for analysis of either unfractionated or fractionated antibodies, for example, during the characterization of fractions isolated from chromatographic techniques such as CEX or SEC, as described earlier. In addition to analysis of undigested antibodies, rpHPLC-MS is a powerful technique that is also routinely applied for the analysis of digested antibodies, whether as part of peptide map analysis or through subunit analysis by limited proteolytic methods (Sjögren, Olsson, & Beck, 2016).

5.6 Affinity-based separation methods

Affinity chromatography can be a powerful separation technique that allows capture and purification of specific analytes from mixtures, including complex solutions such as cell-culture fluid, though specific binding to a unique domain on the analyte of interest. This approach can be used to both separate and concentrate the analyte of interest. The most obvious affinity-based method used with monoclonal antibodies is the purification of antibodies using protein A chromatography (Hober, Nord, & Linhult, 2007). In an analytical context, affinity separation methods such as protein A or protein L are also commonly used, primarily for product concentration determinations, but may also have utility for the analysis of purified peaks obtained by other methods or to separate specific antibody sub-domains, for example, when purification of fragmented Fc or antigen-binding domain fractions is required.

Boronate affinity chromatography (BAC) can be used to selectively bind proteins containing glycation (Brady, Martinez, & Balland, 2007), which is commonly observed on monoclonal antibodies due to the inclusion of reducing sugars such as glucose or mannose during cell-culture production or may occur upon product storage when products are stored in sucrose-containing formulations due to sucrose degradation to reducing sugars (Banks et al., 2009). In this technique, molecules containing glycated residues, typically surface-exposed lysine amino acids distributed across the antibody sequence, are bound to the boronate resin in a weak buffer, promoting interaction between the diol groups of the glycated species and the boronic acid moieties of the resin. Unbound molecules are not retained and eluted in the flow-through. The bound antibody species are eluted using a step or gradient elution of a displacer, such as sorbitol, that disrupts the diol interaction of the analyte. Total degree of glycation can be quantified by integration of all observed protein peaks by a relative percentage calculation. Detection is typically performed by UV absorbance in the range of 214 or 280 nm.

In addition to analytical affinity-based separation methods such as boronate and protein A, specialized modes of separation also exist that can be used for functional characterization of antibodies through immobilization of Fc-receptors such as FcRn (Schlothauer et al., 2013) or FcGammaRIIIa (Kiyoshi et al., 2018). These approaches can be used to reveal subpopulations of antibodies with differential properties related to biological function and are thus of a high interest for understanding and classifying antibody CQAs related to receptor binding, such as Fc-domain N-linked glycosylation (Brady et al., 2015). Alternatively, affinity-based separation techniques can be applied to separate antibodies through differential binding of *N*-glycans by use of immobilized lectins such as concanavalin A (Lu, Westland, Ma, & Gadgil, 2012).

5.7 Hyphenation of separation methods and detection by mass spectrometry

Hyphenation in the context of chromatography refers to the use of multiple separation methods in series and is typically used to separate complex mixtures or when orthogonal techniques are operated in tandem. Two-dimensional (2D) chromatography for antibodies is not commonly employed for routine analyses but may be considered in specialized techniques or applications. Protein A capture prior to analytical chromatography may be used to enable product quality screening of low concentration or unpurified samples during early development stages (Sandra, Steenbeke, Vandenheede, Vanhoenacker, & Sandra, 2017). Alternatively, 2D-LC techniques may enable hyphenation of salt-based separation techniques that cannot typically be directly interfaced with mass spectrometric detectors (Stoll, Danforth, Zhang, & Beck, 2016).

Although MS is not the scope of this chapter, it is worth mentioning the value of interfacing separation techniques with this detection mode, as it is a key aspect of characterization efforts for monoclonal antibodies. The use of rpHPLC in tandem with MS is one the most ubiquitous kinds of method hyphenation techniques in use in the biotechnology industry (Dillon et al., 2006).

As mentioned in the previous section, protein A affinity may be used to purify antibodies or antibody fragments containing Fc domains. This method may also be used as an analytical technique in conjunction with MS (Prentice, Wallace, & Eakin, 2015), typically for desalting of antibodies during characterization studies prior to analysis by MS. This approach is often used when rapid analysis is desired and separation of the antibody into different components (e.g., after reduction) is not required. Similar to this approach, size-exclusion chromatography may also be used in a similar mode (Liu, Gaza-Bulseco, & Chumsae, 2009). In this mode of separation the sieving nature of the chromatography column is used to separate the buffer components from the antibody and allows rapid desalting of antibody samples or fractions during characterization studies. The selection of rpHPLC, SEC, or (less commonly) protein A as a separation technique for monoclonal antibodies prior to MS is usually based on the particular analytical question under inquiry or, more commonly, based on the preference of the laboratory and personal choice of the analyst. In addition to the more typically employed hyphenation of rpHPLC and desalting separation methods mentioned earlier, recent advances in icIEF methodology and instrument design have allowed direct interfacing with MS, further demonstrating the value of this technique for characterization of biopharmaceuticals and monoclonal antibodies (Mack et al., 2019).

5.8 Conclusion and summary

In this chapter a high-level overview of the main separation methods applied to monoclonal antibodies has been presented. Since the first developed antibody drugs for clinical use during the 1980s, separation methods applied for analysis of these important therapeutics have developed and become more powerful in tandem with developments in technology, which are also outlined in this chapter. As the depth of understanding of the purity variants that can be identified as part of analytical testing has deepened, improvements in the framework of how to classify the criticality of these variants have also developed. Through this process, a refined approach to antibody purity testing using separation-based techniques has been developed and largely standardized across the biotechnology industry. A key development in this process has been the deeper understanding of these methods that can be afforded by the application of orthogonal techniques during product testing. The ability to isolate or generate HMW species as quantified by SEC and assess them by a panel of assays, including both in vitro and in vivo tests, has greatly advanced the understanding across industry for both the composition and the meaningfulness (criticality) of these variants. By applying multiple separation techniques to the same analytical challenge, it has become possible not to simply understand that a certain hypothetical antibody has a specific level of aggregate of a certain molecular weight as determined by light scattering and MS but also that this aggregate is composed of both covalent and noncovalent species as analyzed by CE-SDS, is elevated in basic charge variants, and results in increased biological activity in a functional test of antibody activity. This hypothetical but plausible example demonstrates well both the goal and the power of separations-based analytical testing as applied to monoclonal antibodies over the past three decades.

As the analytical challenges encountered evolve, so must the solutions applied to them adapt, whether it be more challenging complex products such as bispecific antibodies, nonantibody-based fusion proteins, or cell-based therapies or challenges of applications such as high concentration formulations or the use of administration devices such as closed system transfer devices (Petoskey, Kwok, Jackson, & Jiang, 2020). Irrespective of the challenge, the use of separations-based analytical methods for the characterization and routine testing of biotherapeutics will remain a key element of analytical science and will continue to evolve and improve with regard to robustness, capability, and speed.

References

Alt, N., Zhang, T. Y., Motchnik, P., Taticek, R., Quarmby, V., Schlothauer, T., ... Harris, R. J. (2016). Determination of critical quality attributes for monoclonal antibodies using quality by design principles. *Biologicals: Journal of the International Association of Biological Standardization*, *44*(5), 291–305, ISSN 1045–1056. Available from https://doi.org/10.1016/j.biologicals.2016.06.005.

Arakawa, T., Ejima, D., Li, T., & Philo, J. S. (2010). The critical role of mobile phase composition in size exclusion chromatography of protein pharmaceuticals. *Journal of Pharmaceutical Sciences*, *99*(4), 1674–1692. Available from https://doi.org/10.1002/jps.21974, ISSN 0022-3549.

Bailly, M., Mieczkowski, C., Juan, V., Metwally, E., Tomazela, D., Baker, J., ... Yu, Y. (2020). Predicting antibody developability profiles through early stage discovery screening. *Mabs*, *12*(1), 1743053.

Banks, D. D., Hambly, D. M., Scavezze, J. L., Siska, C. C., Stackhouse, N. L., & Gadgil, H. S. (2009). The effect of sucrose hydrolysis on the stability of protein therapeutics during accelerated formulation studies. *Journal of Pharmaceutical Sciences*, *98*(12), 4501–4510.

Battersby, R. V., Ohlrogge, B., Feigler, A., & Holloway, C. J. (1992). Isoforms of recombinant human erythropoietin. In H. Pagel, C. Weiss, & W. Jelkmann (Eds.), *Pathophysiology and pharmacology of erythropoietin*. Berlin, Heidelberg: Springer, https://doi.org/10.1007/978-3-642-77074-6_2.

Bessa, J., Boeckle, S., Beck, H., et al. (2015). The immunogenicity of antibody aggregates in a novel transgenic mouse model. *Pharmaceutical Research*, *32*, 2344–2359. Available from https://doi.org/10.1007/s11095-015-1627-0.

Bickel, F., Herold, E. M., Signes, A., Romeijn, S., Jiskoot, W., & Kiefer, H. (2016). Reversible NaCl-induced aggregation of a monoclonal antibody at low pH: Characterization of aggregates and factors affecting aggregation. *European Journal of Pharmaceutics and Biopharmaceutics*, *107*, 310–320. Available from https://doi.org/10.1016/j.ejpb.2016.07.020, ISSN 0939-6411.

Boll, B., Josse, L., Heubach, A., et al. (2018). Impact of non-ideal analyte behavior on the separation of protein aggregates by asymmetric flow field-flow fractionation. *Journal of Separation Science*, *41*, 2854–2864. Available from https://doi.org/10.1002/jssc.201701457.

Boyd, D., Ebrahimi, A., Ronan, S., Mickus, B., Schenauer, M., Wang, J., ... Ambrogelly, A. (2018). Isolation and characterization of a monoclonal antibody containing an extra heavy-light chain Fab arm. *MAbs*, *10*(3), 346–353.

Boyd, D., Kaschak, T., & Yan, B. (2011). HIC resolution of an IgG1 with an oxidized Trp in a complementarity determining region. *Journal of Chromatography B*, *879* (13–14), 955–960.

Brady, L. J., Martinez, T., & Balland, A. (2007). Characterization of nonenzymatic glycation on a monoclonal antibody. *Analytical Chemistry*, *79*(24), 9403–9413.

Brady, L. J., Velayudhan, J., Visone, D. B., Daugherty, K. C., Bartron, J. L., Coon, M., ... Connell-Crowley, L. (2015). The criticality of high-resolution N-linked carbohydrate assays and detailed characterization of antibody effector function in the context of biosimilar development. *MAbs*, 7(3), 562–570.

Brock, D. C., & Moore, G. E. (Eds.), (2006). *Understanding Moore's law: Four decades of innovation*. Chemical Heritage Foundation.

Brusotti, G., Calleri, E., Colombo, R., Massolini, G., Rinaldi, F., & Temporini, C. (2018). Advances on size exclusion chromatography and applications on the analysis of protein biopharmaceuticals and protein aggregates: A mini review. *Chromatographia, 81*(1), 3–23.

Brych, S. R., Gokarn, Y. R., Hultgen, H., Stevenson, R. J., Rajan, R., & Matsumura, M. (2010). Characterization of antibody aggregation: Role of buried, unpaired cysteines in particle formation. *Journal of Pharmaceutical Sciences, 99*(2), 764–781. Available from https://doi.org/10.1002/jps.21868, ISSN 0022-3549.

Cai, B., Pan, H., & Flynn, G. C. (2011). C-terminal lysine processing of human immunoglobulin G2 heavy chain in vivo. *Biotechnology and Bioengineering, 108*, 404–412. Available from https://doi.org/10.1002/bit.22933.

Cao, M., Mulagapati, S. H. R., Vemulapalli, B., Wang, J., Saveliev, S. V., Urh, M., ... Liu, D. (2019). Characterization and quantification of succinimide using peptide mapping under low-pH conditions and hydrophobic interaction chromatography. *Analytical Biochemistry, 566*, 151–159.

Chen, G., Ha, S., & Rustandi, R. R. (2013). *Characterization of glycoprotein biopharmaceutical products by Caliper LC90 CE-SDS gel technology. Glycosylation engineering of biopharmaceuticals* (pp. 199–209). Totowa, NJ: Humana Press.

Chen, X., Tang, K., Lee, M., & Flynn, G. C. (2008). Microchip assays for screening monoclonal antibody product quality. *Electrophoresis, 29*(24), 4993–5002.

Chiu, M. L., Goulet, D. R., Teplyakov, A., & Gilliland, G. L. (2019). Antibody structure and function: The basis for engineering therapeutics. *Antibodies (Basel), 8*(4), 55. Available from http://doi.org/10.3390/antib8040055.

Couderc, F., Ong-Meang, V., & Poinsot, V. (2017). Capillary electrophoresis hyphenated with UV-native-laser induced fluorescence detection (CE/UV-native-LIF). *Electrophoresis, 38*, 135–149. Available from https://doi.org/10.1002/elps.201600248.

Cromwell, M. E. M., Hilario, E., & Jacobson, F. (2006). Protein aggregation and bioprocessing. *The AAPS Journal, 8*, E572–E579. Available from https://doi.org/10.1208/aapsj080366.

Dada, O. O., Jaya, N., Valliere-Douglass, J., & Salas-Solano, O. (2015). Characterization of acidic and basic variants of IgG1 therapeutic monoclonal antibodies based on non-denaturing IEF fractionation. *Electrophoresis, 36*, 2695–2702. Available from https://doi.org/10.1002/elps.201500219.

Diaz, S. L., Padler-Karavani, V., Ghaderi, D., Hurtado-Ziola, N., Yu, H., Chen, X., et al. (2009). Sensitive and specific detection of the non-human sialic acid *N*-glycolyl neuraminic acid in human tissues and biotherapeutic products. *PLoS One, 4*(1), e4241. Available from https://doi.org/10.1371/journal.pone.0004241.

Dillon, T. M., Bondarenko, P. V., Rehder, D. S., Pipes, G. D., Kleemann, G. R., & Ricci, M. S. (2006). Optimization of a reversed-phase high-performance liquid chromatography/mass spectrometry method for characterizing recombinant antibody heterogeneity and stability. *Journal of Chromatography A, 1120*(1–2), 112–120.

Doss, H. R., Raman, M., Knihtila, R., Chennamsetty, N., Wang, D., Shupe, A., & Mussa, N. (2019). Streamlining the polishing step development process via physicochemical characterization of monoclonal antibody aggregates. *Journal of Chromatography A, 1598*, 101–112. Available from https://doi.org/10.1016/j.chroma.2019.03.044, ISSN 0021-9673.

Ejima, D., Yumioka, R., Arakawa, T., & Tsumoto, K. (2005). Arginine as an effective additive in gel permeation chromatography. *Journal of Chromatography A, 1094*, 49–55.

Esfandiary, R., Hayes, D. B., Parupudi, A., Casas-Finet, J., Bai, S., Samra, H. S., ... Sathish, H. A. (2013). A systematic multitechnique approach for detection and characterization of reversible self-association during formulation development of therapeutic antibodies. *Journal of Pharmaceutical Sciences*, *102*(9), 3089–3099. Available from https://doi.org/10.1002/jps.23654, ISSN 0022-3549.

Esterman, A. L., Katiyar, A., & Krishnamurthy, G. (2016). Implementation of USP antibody standard for system suitability in capillary electrophoresis sodium dodecyl sulfate (CE-SDS) for release and stability methods. *Journal of Pharmaceutical and Biomedical Analysis*, *128*, 447–454.

Farsang, E., Murisier, A., Horváth, K., Beck, A., Kormány, R., Guillarme, D., & Fekete, S. (2019). Tuning selectivity in cation-exchange chromatography applied for monoclonal antibody separations, part 1: Alternative mobile phases and fine tuning of the separation. *Journal of Pharmaceutical and Biomedical Analysis*, *168*, 138–147. Available from https://doi.org/10.1016/j.jpba.2019.02.024, ISSN 0731-7085.

Fathallah, A. M., Chiang, M., Mishra, A., Kumar, S., Xue, L., Middaugh, C. R., & Balu-Iyer, S. V. (2015). The effect of small oligomeric protein aggregates on the immunogenicity of intravenous and subcutaneous administered antibodies. *Journal of Pharmaceutical Sciences*, *104*(11), 3691–3702.

Fekete, S., Beck, A., Fekete, J., & Guillarme, D. (2015a). Method development for the separation of monoclonal antibody charge variants in cation exchange chromatography, Part I: Salt gradient approach. *Journal of Pharmaceutical and Biomedical Analysis*, *102*, 33–44, ISSN 0731-7085. Available from https://doi.org/10.1016/j.jpba.2014.08.035.

Fekete, S., Beck, A., Fekete, J., & Guillarme, D. (2015b). Method development for the separation of monoclonal antibody charge variants in cation exchange chromatography, Part II: pH gradient approach. *Journal of Pharmaceutical and Biomedical Analysis*, *102*, 282–289. Available from https://doi.org/10.1016/j.jpba.2014.09.032.

Fekete, S., Gassner, A.-L., Rudaz, S., Schappler, J., & Guillarme, D. (2013). Analytical strategies for the characterization of therapeutic monoclonal antibodies. *TrAC Trends in Analytical Chemistry*, *42*, 74–83. Available from https://doi.org/10.1016/j.trac.2012.09.012, ISSN 0165-9936.

Fekete, S., Veuthey, J.-L., & Guillarme, D. (2017). Achievable separation performance and analysis time in current liquid chromatographic practice for monoclonal antibody separations. *Journal of Pharmaceutical and Biomedical Analysis*, *141*, 59–69. Available from https://doi.org/10.1016/j.jpba.2017.04.004, ISSN 0731-7085.

Filipe, V., Jiskoot, W., Basmeleh, A. H., Halim, A., Schellekens, H., & Brinks, V. (2012). Immunogenicity of different stressed IgG monoclonal antibody formulations in immune tolerant transgenic mice. *mAbs*, *4*(6), 740–752. Available from https://doi.org/10.4161/mabs.22066.

Finkler, C., & Krummen, L. (2016). Introduction to the application of QbD principles for the development of monoclonal antibodies. *Biologicals: Journal of the International Association of Biological Standardization*, *44*(5), 282–290.

Flatman, S., Alam, I., Gerard, J., & Mussa, N. (2007). Process analytics for purification of monoclonal antibodies. *Journal of Chromatography B*, *848*(1), 79–87. Available from https://doi.org/10.1016/j.jchromb.2006.11.018.

Gabrielson, J. P., Brader, M. L., Pekar, A. H., Mathis, K. B., Winter, G., Carpenter, J. F., & Randolph, T. W. (2007). Quantitation of aggregate levels in a recombinant humanized monoclonal antibody formulation by size-exclusion chromatography, asymmetrical flow field flow fractionation, and sedimentation velocity. *Journal of Pharmaceutical Sciences*, *96*(2), 268–279. Available from https://doi.org/10.1002/jps.20760, ISSN 0022-3549.

Gandhi, S., Ren, D., Xiao, G., et al. (2012). Elucidation of degradants in acidic peak of cation exchange chromatography in an IgG1 monoclonal antibody formed on long-term storage in a liquid formulation. *Pharmaceutical Research*, *29*, 209–224. Available from https://doi.org/10.1007/s11095-011-0536-0.

Ghrist, B. F. D., Stadalius, M. A., & Snyder, L. R. (1987). Predicting bandwidth in the high-performance liquid chromatographic separation of large biomolecules: I. Size-exclusion studies and the role of solute stokes diameter versus particle pore diameter. *Journal of Chromatography A*, *387*, 1–19. Available from https://doi.org/10.1016/S0021-9673(01)94510-8, ISSN 0021-9673.

Goetze, A. M., Schenauer, M. R., & Flynn, G. C. (2010). Assessing monoclonal antibody product quality attribute criticality through clinical studies. *mAbs*, *2*(5), 500–507. Available from https://doi.org/10.4161/mabs.2.5.12897.

Gomez, N., Subramanian, J., Ouyang, J., Nguyen, M. D., Hutchinson, M., Sharma, V. K., ... Yuk, I. H. (2012). Culture temperature modulates aggregation of recombinant antibody in CHO cells. *Biotechnology and Bioengineering*, *109*, 125–136. Available from https://doi.org/10.1002/bit.23288.

Goyon, A., Beck, A., Colas, O., Sandra, K., Guillarme, D., & Fekete, S. (2017). Evaluation of size exclusion chromatography columns packed with sub-3μm particles for the analysis of biopharmaceutical proteins. *Journal of Chromatography A*, *1498*, 80–89. Available from https://doi.org/10.1016/j.chroma.2016.11.056, ISSN 0021-9673.

Goyon, A., Beck, A., Veuthey, J. L., Guillarme, D., & Fekete, S. (2017). Comprehensive study on the effects of sodium and potassium additives in size exclusion chromatographic separations of protein biopharmaceuticals. *Journal of Pharmaceutical and Biomedical Analysis*, *144*, 242–251. Available from https://doi.org/10.1016/j.jpba.2016.09.031, Epub 2016 Sep 26. PMID: 27697310.

Goyon, A., Excoffier, M., Janin-Bussat, M.-C., Bobaly, B., Fekete, S., Guillarme, D., & Beck, A. (2017). Determination of isoelectric points and relative charge variants of 23 therapeutic monoclonal antibodies. *Journal of Chromatography B*, *1065–1066*, 119–128. Available from https://doi.org/10.1016/j.jchromb.2017.09.033, ISSN 1570-0232.

Goyon, A., Guillarme, D., & Fekete, S. (2017). The importance of system band broadening in modern size exclusion chromatography. *Journal of Pharmaceutical and Biomedical Analysis*, *135*, 50–60. Available from https://doi.org/10.1016/j.jpba.2016.12.004, ISSN 0731-7085.

Harris, C., Xu, W., Grassi, L., Wang, C., Markle, A., Hardman, C., ... Wang, J. (2019). Identification and characterization of an IgG sequence variant with an 11 kDa heavy chain C-terminal extension using a combination of mass spectrometry and high-throughput sequencing analysis. *MAbs*, *11*(8), 1452–1463.

Harris, R. J., Kabakoff, B., Macchi, F. D., Shen, F. J., Kwong, M., Andya, J. D., ... Chen, A. B. (2001). Identification of multiple sources of charge heterogeneity in a recombinant antibody. *Journal of Chromatography B: Biomedical Sciences and Applications*, *752*(2), 233–245. Available from https://doi.org/10.1016/S0378-4347(00)00548-X, ISSN 0378-4347.

Hober, S., Nord, K., & Linhult, M. (2007). Protein A chromatography for antibody purification. *Journal of Chromatography B*, *848*(1), 40–47.

Hong, P., Koza, S., & Bouvier, E. S. (2012). A review size-exclusion chromatography for the analysis of protein biotherapeutics and their aggregates. *Journal of Liquid Chromatography & Related Technologies*, *35*(20), 2923–2950.

Hosken, B. D., Li, C., Mullappally, B., Co, C., & Zhang, B. (2016). Isolation and characterization of monoclonal antibody charge variants by free flow isoelectric focusing. *Analytical Chemistry*, *88*(11), 5662–5669. Available from https://doi.org/10.1021/acs.analchem.5b03946.

Javeri, I., Nellaiappan, K., McNemar, C., Yakovlevsky, K., Soukrati, A., Velisetty, P., & Misra, B. (2018). Use of cyclodextrin as a novel agent in the SEC-HPLC mobile phase to mitigate the interactions of proteins or peptide or their impurities with the residual silanols of commercial SEC-HPLC columns with improved separation and resolution. *Pharmaceutical Research, 35*(9), 1−17.

Jin, W., Xing, Z., Song, Y., Huang, C., Xu, X., Ghose, S., & Li, Z. J. (2019). Protein aggregation and mitigation strategy in low pH viral inactivation for monoclonal antibody purification. *mAbs, 11*(8), 1479−1491. Available from https://doi.org/10.1080/19420862.2019.1658493.

Jooß, K., Hühner, J., Kiessig, S., et al. (2017). Two-dimensional capillary zone electrophoresis−mass spectrometry for the characterization of intact monoclonal antibody charge variants, including deamidation products. *Analytical and Bioanalytical Chemistry, 409*, 6057−6067. Available from https://doi.org/10.1007/s00216-017-0542-0.

Joshi, V. S., Kumar, V., & Rathore, A. S. (2017). Optimization of ion exchange sigmoidal gradients using hybrid models: Implementation of quality by design in analytical method development. *Journal of Chromatography A, 1491*, 145−152. Available from https://doi.org/10.1016/j.chroma.2017.02.058, ISSN 0021-9673.

Joubert, M. K., Luo, Q., Nashed-Samuel, Y., Wypych, J., & Narhi, L. O. (2011). Classification and characterization of therapeutic antibody aggregates. *Journal of Biological Chemistry, 286*(28), 25118−25133. Available from https://doi.org/10.1074/jbc.M110.160457.

Kahle, J., Maul, K. J., & Wätzig, H. (2018). The next generation of capillary electrophoresis instruments: Performance of CE-SDS protein analysis. *Electrophoresis, 39*, 311−325. Available from https://doi.org/10.1002/elps.201700278.

Kahle, J., Stein, M., & Wätzig, H. (2019). Design of experiments as a valuable tool for biopharmaceutical analysis with (imaged) capillary isoelectric focusing. *Electrophoresis, 40*(18−19), 2382−2389.

Kahle, J., & Wätzig, H. (2018). Determination of protein charge variants with (imaged) capillary isoelectric focusing and capillary zone electrophoresis. *Electrophoresis, 39*, 2492−2511. Available from https://doi.org/10.1002/elps.201800079.

Kahle, J., Zagst, H., Wiesner, R., & Wätzig, H. (2019). Comparative charge-based separation study with various capillary electrophoresis (CE) modes and cation exchange chromatography (CEX) for the analysis of monoclonal antibodies. *Journal of Pharmaceutical and Biomedical Analysis, 174*, 460−470. Available from https://doi.org/10.1016/j.jpba.2019.05.058, ISSN 0731-7085.

Kaschak, T., Boyd, D., Lu, F., Derfus, G., Kluck, B., Nogal, B., ... Yan, B. (2011). Characterization of the basic charge variants of a human IgG1. *mAbs, 3*(6), 577−583. Available from https://doi.org/10.4161/mabs.3.6.17959.

Kaschak, T., Boyd, D., & Yan, B. (2011). Characterization of glycation in an IgG1 by capillary electrophoresis sodium dodecyl sulfate and mass spectrometry. *Analytical Biochemistry, 417*(2), 256−263.

Kepert, J. F., Cromwell, M., Engler, N., Finkler, C., Gellermann, G., Gennaro, L., ... Taticek, R. (2016). Establishing a control system using QbD principles. *Biologicals: Journal of the International Association of Biological Standardization, 44*(5), 319−331. Available from https://doi.org/10.1016/j.biologicals.2016.06.003, ISSN 1045-1056.

Kijanka, G., Bee, J. S., Bishop, S. M., Que, I., Löwik, C., & Jiskoot, W. (2016). Fate of multimeric oligomers, submicron, and micron size aggregates of monoclonal antibodies upon subcutaneous injection in mice. *Journal of Pharmaceutical Sciences, 105*(5), 1693−1704.

Kiyoshi, M., Caaveiro, J. M., Tada, M., Tamura, H., Tanaka, T., Terao, Y., ... Kuroda, D. (2018). Assessing the heterogeneity of the Fc-glycan of a therapeutic antibody using an engineered Fcγ receptor IIIa-immobilized column. *Scientific Reports, 8*(1), 1−11.

Kochling, J., Wu, W., Hua, Y., Guan, Q., & Castaneda-Merced, J. (2016). A platform analytical quality by design (AQbD) approach for multiple UHPLC-UV and

UHPLC–MS methods development for protein analysis. *Journal of Pharmaceutical and Biomedical Analysis, 125*, 130–139. Available from https://doi.org/10.1016/j.jpba.2016.03.031, ISSN 0731-7085.

Kohler, G., & Milstein, C. (1975). Continuous cultures of fused cells secreting antibody–antigen interactions. *Nature, 256*(5517), 495–497.

Kükrer, B., Filipe, V., van Duijn, E., et al. (2010). Mass spectrometric analysis of intact human monoclonal antibody aggregates fractionated by size-exclusion chromatography. *Pharmaceutical Research, 27*, 2197–2204. Available from https://doi.org/10.1007/s11095-010-0224-5.

Lacher, N. A., Roberts, R. K., He, Y., Cargill, H., Kearns, K. M., Holovics, H., & Ruesch, M. N. (2010). Development, validation, and implementation of capillary gel electrophoresis as a replacement for SDS-PAGE for purity analysis of IgG2 mAbs. *Journal of Separation Science, 33*, 218–227. Available from https://doi.org/10.1002/jssc.200900597.

Lacher, N. A., Wang, Q., Roberts, R. K., Holovics, H. J., Aykent, S., Schlittler, M. R., ... Demarest, C. W. (2010). Development of a capillary gel electrophoresis method for monitoring disulfide isomer heterogeneity in IgG2 antibodies. *Electrophoresis, 31*(3), 448–458.

Lee, J., Kang, H. A., Bae, J. S., Kim, K. D., Lee, K. H., Lim, K. J., ... Chang, S. J. (2018). Evaluation of analytical similarity between trastuzumab biosimilar CT-P6 and reference product using statistical analyses. *mAbs, 10*(4), 547–571. Available from https://doi.org/10.1080/19420862.2018.1440170.

Lindqvist, B., & Storgårds, T. (1955). Molecular-sieving properties of starch. *Nature, 175* (4455), 511–512.

Li, W., Yang, B., Zhou, D., Xu, J., Li, W., & Suen, W.-C. (2017). Identification and characterization of monoclonal antibody fragments cleaved at the complementarity determining region using orthogonal analytical methods. *Journal of Chromatography B, 1048*, 121–129. Available from https://doi.org/10.1016/j.jchromb.2017.02.019, ISSN 1570-0232.

Li, Y., Huang, Y., Ferrant, J., Lyubarskaya, Y., (Emma) Zhang, Y., (Peter) Li, S., & (Billy) Wu, S.-L. (2016). Assessing in vivo dynamics of multiple quality attributes from a therapeutic IgG4 monoclonal antibody circulating in cynomolgus monkey. *mAbs, 8* (5), 961–968. Available from https://doi.org/10.1080/19420862.2016.1167298.

Lingg, N., Berndtsson, M., Hintersteiner, B., Schuster, M., Bardor, M., & Jungbauer, A. (2014). Highly linear pH gradients for analyzing monoclonal antibody charge heterogeneity in the alkaline range: Validation of the method parameters. *Journal of Chromatography A, 1373*, 124–130. Available from https://doi.org/10.1016/j.chroma.2014.11.021, ISSN 0021-9673.

Lingg, N., Tan, E., Hintersteiner, B., Bardor, M., & Jungbauer, A. (2013). Highly linear pH gradients for analyzing monoclonal antibody charge heterogeneity in the alkaline range. *Journal of Chromatography A, 1319*, 65–71, ISSN 0021-9673. Available from https://doi.org/10.1016/j.chroma.2013.10.028.

Liu, H., Gaza-Bulseco, G., & Chumsae, C. (2009). Analysis of reduced monoclonal antibodies using size exclusion chromatography coupled with mass spectrometry. *Journal of the American Society for Mass Spectrometry, 20*(12), 2258–2264.

Liu, J., Nguyen, M. D. H., Andya, J. D., & Shire, S. J. (2005). Reversible self-association increases the viscosity of a concentrated monoclonal antibody in aqueous solution. *Journal of Pharmaceutical Sciences, 94*(9), 1928–1940. Available from https://doi.org/10.1002/jps.20347, ISSN 0022-3549.

Liu, S., Madren, S., Feng, P., & Sosic, Z. (2020). Characterization of the acidic species of a monoclonal antibody using free flow electrophoresis fractionation and mass spectrometry. *Journal of Pharmaceutical and Biomedical Analysis, 185*, 113217. Available from https://doi.org/10.1016/j.jpba.2020.113217.

Luo, S., McSweeney, K. M., Wang, T., Bacot, S. M., Feldman, G. M., & Zhang, B. (2020). Defining the right diluent for intravenous infusion of therapeutic antibodies. *Mabs, 12*(1), 1685814.

Lu, Y., Westland, K., Ma, Y. H., & Gadgil, H. (2012). Evaluation of effects of Fc domain high-mannose glycan on antibody stability. *Journal of Pharmaceutical Sciences, 101*(11), 4107–4117.

Mack, S., Arnold, D., Bogdan, G., Bousse, L., Danan, L., Dolnik, V., ... Ji, J. (2019). A novel microchip-based imaged CIEF-MS system for comprehensive characterization and identification of biopharmaceutical charge variants. *Electrophoresis, 40*(23–24), 3084–3091.

Mack, S., & Darling, S. (2014). Mastering charge heterogeneity analysis of therapeutic proteins: Understanding crucial aspects of icIEF method development and implementation. *Genetic Engineering & Biotechnology News, 34*(7), 38–39.

Martínez-Ortega, A., Herrera, A., Salmerón-García, A., Cabeza, J., Cuadros-Rodríguez, L., & Navas, N. (2018). Validated reverse phase HPLC diode array method for the quantification of intact bevacizumab, infliximab and trastuzumab for long-term stability study. *International Journal of Biological Macromolecules, 116*, 993–1003.

Meert, C. D., Brady, L. J., Guo, A., & Balland, A. (2010). Characterization of antibody charge heterogeneity resolved by preparative immobilized pH gradients. *Analytical Chemistry, 82*(9), 3510–3518. Available from https://doi.org/10.1021/ac902408r.

Miao, S., Xie, P., Zou, M., et al. (2017). Identification of multiple sources of the acidic charge variants in an IgG1 monoclonal antibody. *Applied Microbiology and Biotechnology, 101*, 5627–5638. Available from https://doi.org/10.1007/s00253-017-8301-x.

Michels, D. A., Ip, A. Y., Dillon, T. M., Brorson, K., Lute, S., Chavez, B., ... and Miller, K. J. (2015). State-of-the-art and emerging technologies for therapeutic monoclonal antibody characterization. In: *Volume 2. Biopharmaceutical characterization: The NISTmAb case study* (Chapter 5; pp. 237–284). ACS Publications.

Michels, D. A., Brady, L. J., Guo, A., & Balland, A. (2007). Fluorescent derivatization method of proteins for characterization by capillary electrophoresis–sodium dodecyl sulfate with laser-induced fluorescence detection. *Analytical Chemistry, 79*(15), 5963–5971. Available from https://doi.org/10.1021/ac0705521.

Michels, D. A., Parker, M., & Salas-Solano, O. (2012). Quantitative impurity analysis of monoclonal antibody size heterogeneity by CE-LIF: Example of development and validation through a quality-by-design framework. *Electrophoresis, 33*, 815–826. Available from https://doi.org/10.1002/elps.201100528.

Molina, P., Schick, A. J., Welch, L., Niedringhaus, T., del Hierro, G., Deperalta, G., & Hieb, A. (2020). Using differential scanning calorimetry for the development of non-reduced capillary electrophoresis sodium dodecyl sulfate methods for monoclonal antibodies. *Analytical Biochemistry, 609*, 113948. Available from https://doi.org/10.1016/j.ab.2020.113948.

Morbioli, G. G., Mazzu-Nascimento, T., Aquino, A., Cervantes, C., & Carrilho, E. (2016). Recombinant drugs-on-a-chip: The usage of capillary electrophoresis and trends in miniaturized systems—A review. *Analytica Chimica Acta, 935*, 44–57.

Moritz, B., Locatelli, V., Niess, M., Bathke, A., Kiessig, S., Entler, B., ... Stracke, J. (2017). Optimization of capillary zone electrophoresis for charge heterogeneity testing of biopharmaceuticals using enhanced method development principles. *Electrophoresis, 38*, 3136–3146. Available from https://doi.org/10.1002/elps.201700145.

Moritz, B., Schnaible, V., Kiessig, S., Heyne, A., Wild, M., Finkler, C., ... Lies, M. (2015). Evaluation of capillary zone electrophoresis for charge heterogeneity testing of monoclonal antibodies. *Journal of Chromatography B, 983–984*, 101–110. Available from https://doi.org/10.1016/j.jchromb.2014.12.024, ISSN 1570-0232.

Moritz, B., & Stracke, J. O. (2017). Assessment of disulfide and hinge modifications in monoclonal antibodies. *Electrophoresis, 38*(6), 769–785.

Mou, X., Yang, X., Li, H., Ambrogelly, A., & Pollard, D. J. (2014). A high throughput ultra performance size exclusion chromatography assay for the analysis of aggregates and fragments of monoclonal antibodies. *Pharmaceutical Bioprocessing, 2*(2), 141–156.

Nicoud, L., Jagielski, J., Pfister, D., Lazzari, S., Massant, J., Lattuada, M., & Morbidelli, M. (2016). Kinetics of monoclonal antibody aggregation from dilute toward concentrated conditions. *The Journal of Physical Chemistry B, 120*(13), 3267–3280. Available from https://doi.org/10.1021/acs.jpcb.5b11791.

Perkins, M., Theiler, R., Lunte, S., et al. (2000). Determination of the origin of charge heterogeneity in a murine monoclonal antibody. *Pharmaceutical Research, 17*, 1110–1117. Available from https://doi.org/10.1023/A:1026461830617.

Petoskey, F., Kwok, S. C., Jackson, W., & Jiang, S. (2020). Overcoming challenges of implementing closed system transfer device clinical in-use compatibility testing for drug development of antibody drug conjugates. *Journal of Pharmaceutical Sciences, 109*(1), 761–768.

Prentice, K. M., Wallace, A., & Eakin, C. M. (2015). Inline protein A mass spectrometry for characterization of monoclonal antibodies. *Analytical Chemistry, 87*(4), 2023–2028.

Quan, C., Alcala, E., Petkovska, I., Matthews, D., Canova-Davis, E., Taticek, R., & Ma, S. (2008). A study in glycation of a therapeutic recombinant humanized monoclonal antibody: Where it is, how it got there, and how it affects charge-based behavior. *Analytical Biochemistry, 373*(2), 179–191. Available from https://doi.org/10.1016/j.ab.2007.09.027, ISSN 0003-2697.

Ren, D., Zhang, J., Pritchett, R., Liu, H., Kyauk, J., Luo, J., & Amanullah, A. (2011). Detection and identification of a serine to arginine sequence variant in a therapeutic monoclonal antibody. *Journal of Chromatography B, 879*(27), 2877–2884. Available from https://doi.org/10.1016/j.jchromb.2011.08.015, ISSN 1570-0232.

Rein, P., & Mueller, R. B. (2017). Treatment with biologicals in rheumatoid arthritis: An overview. *Rheumatology and Therapy, 4*, 247–261. Available from https://doi.org/10.1007/s40744-017-0073-3.

Roumeliotis, P., & Unger, K. K. (1981). Assessment and optimization of system parameters in size exclusion separation of proteins on diol-modified silica columns. *Journal of Chromatography A, 218*, 535–546. Available from https://doi.org/10.1016/S0021-9673(00)82078-6, ISSN 0021-9673.

Ruesch, M. N., Benetti, L., Berkay, E., Cirelli, D. J., Frantz, N., Gastens, M. H., ... Stults, J. (2020). Strategies for setting patient-centric commercial specifications for biotherapeutic products. *Journal of Pharmaceutical Sciences, 110*. Available from https://doi.org/10.1016/j.xphs.2020.09.048, ISSN 0022-3549.

Rustandi, R. R., Washabaugh, M. W., & Wang, Y. (2008). Applications of CE SDS gel in development of biopharmaceutical antibody-based products. *Electrophoresis, 29*, 3612–3620. Available from https://doi.org/10.1002/elps.200700958.

Salas-Solano, O., Babu, K., Park, S. S., et al. (2011). Intercompany study to evaluate the robustness of capillary isoelectric focusing technology for the analysis of monoclonal antibodies. *Chromatographia, 73*, 1137–1144. Available from https://doi.org/10.1007/s10337-011-2017-3.

Salas-Solano, O., Tomlinson, B., Du, S., Parker, M., Strahan, A., & Ma, S. (2006). Optimization and validation of a quantitative capillary electrophoresis sodium dodecyl sulfate method for quality control and stability monitoring of monoclonal antibodies. *Analytical Chemistry, 78*(18), 6583–6594.

Sandra, K., Steenbeke, M., Vandenheede, I., Vanhoenacker, G., & Sandra, P. (2017). The versatility of heart-cutting and comprehensive two-dimensional liquid chromatography in monoclonal antibody clone selection. *Journal of Chromatography A, 1523*, 283–292.

Sänger–van de Griend, C. E. (2019). CE-SDS method development, validation, and best practice—An overview. *Electrophoresis, 40*(18–19), 2361–2374.

Schlothauer, T., Rueger, P., Stracke, J. O., Hertenberger, H., Fingas, F., Kling, L., . . . Koch, S. (2013). Analytical FcRn affinity chromatography for functional characterization of monoclonal antibodies. *MAbs*, *5*(4), 576–586.

Scott, R. A., Rogers, R., Balland, A., & Brady, L. J. (2014). Rapid identification of an antibody DNA construct rearrangement sequence variant by mass spectrometry. *mAbs*, *6*(6), 1453–1463. Available from https://doi.org/10.4161/mabs.36222.

Sheeley, D. M., Merrill, B. M., & Taylor, L. C. E. (1997). Characterization of monoclonal antibody glycosylation: Comparison of expression systems and identification of terminal α-linked galactose. *Analytical Biochemistry*, *247*(1), 102–110. Available from https://doi.org/10.1006/abio.1997.2036, ISSN 0003-2697.

Shi, Y., Li, Z., Qiao, Y., & Lin, J. (2012). Development and validation of a rapid capillary zone electrophoresis method for determining charge variants of mAb. *Journal of Chromatography B*, *906*, 63–68. Available from https://doi.org/10.1016/j.jchromb.2012.08.022.

Shi, R. L., Xiao, G., Dillon, T. M., Ricci, M. S., & Bondarenko, P. V. (2020). Characterization of therapeutic proteins by cation exchange chromatography-mass spectrometry and top-down analysis. *mAbs*, *12*, 1. Available from https://doi.org/10.1080/19420862.2020.1739825.

Shukla, A. A., Hubbard, B., Tressel, T., Guhan, S., & Low, D. (2007). Downstream processing of monoclonal antibodies—Application of platform approaches. *Journal of Chromatography B*, *848*(1), 28–39. Available from https://doi.org/10.1016/j.jchromb.2006.09.026.

Sjögren, J., Olsson, F., & Beck, A. (2016). Rapid and improved characterization of therapeutic antibodies and antibody related products using IdeS digestion and subunit analysis. *Analyst*, *141*(11), 3114–3125.

Smadja, C., Tran, N. T., & Tavernaa, M. (2017). *Capillary electrophoresis for the quality control of intact therapeutic monoclonal antibodies. Capillary electrophoresis: Trends and developments in pharmaceutical research* (p. 305) CRC Press.

Smith, M. T., Zhang, S., Adams, T., DiPaolo, B., & Dally, J. (2017). Establishment and validation of a microfluidic capillary gel electrophoresis platform method for purity analysis of therapeutic monoclonal antibodies. *Electrophoresis*, *38*, 1353–1365. Available from https://doi.org/10.1002/elps.201600519.

Stoll, D., Danforth, J., Zhang, K., & Beck, A. (2016). Characterization of therapeutic antibodies and related products by two-dimensional liquid chromatography coupled with UV absorbance and mass spectrometric detection. *Journal of Chromatography B*, *1032*, 51–60.

Sydow, J. F., Lipsmeier, F., Larraillet, V., Hilger, M., Mautz, B., Mølhøj, M., et al. (2014). Structure-based prediction of asparagine and aspartate degradation sites in antibody variable regions. *PLoS One*, *9*(6), e100736. Available from https://doi.org/10.1371/journal.pone.0100736.

Tang, H., Miao, S., Zhang, X., et al. (2018). Insights into the generation of monoclonal antibody acidic charge variants during Chinese hamster ovary cell cultures. *Applied Microbiology and Biotechnology*, *102*, 1203–1214. Available from https://doi.org/10.1007/s00253-017-8650-5.

Tang, J., Yu, J., Hubbard-Lucey, V., et al. (2018). The clinical trial landscape for PD1/PDL1 immune checkpoint inhibitors. *Nature Reviews Drug Discovery*, *17*, 854–855. Available from https://doi.org/10.1038/nrd.2018.210.

Tebbey, P. W., Varga, A., Naill, M., Clewell, J., & Venema, J. (2015). Consistency of quality attributes for the glycosylated monoclonal antibody Humira® (adalimumab). *mAbs*, 7(5), 805–811. Available from https://doi.org/10.1080/19420862.2015.1073429.

Teshima, G., Li, M.-X., Danishmand, R., Obi, C., To, R., Huang, C., . . . Tomic, M. (2011). Separation of oxidized variants of a monoclonal antibody by anion-exchange. *Journal of Chromatography A*, *1218*(15), 2091–2097. Available from https://doi.org/10.1016/j.chroma.2010.10.107, ISSN 0021-9673.

Thistlethwaite, J. R., Jr, Haag, B. W., Gaber, A. O., Stuart, J. K., Aronson, A. J., Mayes, J. T., ... Stuart, F. P. (1987). The use of OKT3 to treat steroid-resistant renal allograft rejection in patients receiving cyclosporine. *Transplantation Proceedings*, *19*(1 Pt 3), 1901–1904, PMID: 3103294.

Trappe, A., Füssl, F., Carillo, S., Zaborowska, I., Meleady, P., & Bones, J. (2018). Rapid charge variant analysis of monoclonal antibodies to support lead candidate biopharmaceutical development. *Journal of Chromatography B*, *1095*, 166–176. Available from https://doi.org/10.1016/j.jchromb.2018.07.037, ISSN 1570-0232.

Trexler-Schmidt, M., Sargis, S., Chiu, J., Sze-Khoo, S., Mun, M., Kao, Y. H., & Laird, M. W. (2010). Identification and prevention of antibody disulfide bond reduction during cell culture manufacturing. *Biotechnology and Bioengineering*, *106*(3), 452–461.

Tsumoto, K., Ejima, D., Kita, Y., & Arakawa, T. (2005). Review: Why is arginine effective in suppressing aggregation? *Protein and Peptide Letters*, *12*(7), 613–619.

Van Buren, N., Rehder, D., Gadgil, H., Matsumura, M., & Jacob, J. (2009). Elucidation of two major aggregation pathways in an IgG2 antibody. *Journal of Pharmaceutical Sciences*, *98*(9), 3013–3030. Available from https://doi.org/10.1002/jps.21514, ISSN 0022-3549.

Valliere-Douglass, J., Wallace, A., & Balland, A. (2008). Separation of populations of antibody variants by fine tuning of hydrophobic-interaction chromatography operating conditions. *Journal of Chromatography A*, *1214*(1–2), 81–89.

Valliere-Douglass, J. F., Brady, L. J., Farnsworth, C., Pace, D., Balland, A., Wallace, A., ... Yan, B. (2008). *O*-Fucosylation of an antibody light chain: Characterization of a modification occurring on an IgG1 molecule. *Glycobiology*, *19*(2), 144–152.

Valliere-Douglass, J. F., Eakin, C. M., Wallace, A., Ketchem, R. R., Wang, W., Treuheit, M. J., & Balland, A. (2010). Glutamine-linked and non-consensus asparagine-linked oligosaccharides present in human recombinant antibodies define novel protein glycosylation motifs. *Journal of Biological Chemistry*, *285*(21), 16012–16022.

Váradi, C., Jakes, C., & Bones, J. (2020). Analysis of cetuximab *N*-glycosylation using multiple fractionation methods and capillary electrophoresis mass spectrometry. *Journal of Pharmaceutical and Biomedical Analysis*, *180*, 113035.

Vlasak, J., Bussat, M. C., Wang, S., Wagner-Rousset, E., Schaefer, M., Hamour, C. K., ... Beck, A. (2009). Identification and characterization of asparagine deamidation in the light chain CDR1 of a humanized IgG1 antibody. *Analytical Biochemistry*, *392*(2), 145–154. Available from https://doi.org/10.1016/j.ab.2009.05.043, ISSN 0003-2697.

Wagner, E., Colas, O., Chenu, S., Goyon, A., Murisier, A., Cianferani, S., ... Beck, A. (2020). Determination of size variants by CE-SDS for approved therapeutic antibodies: Key implications of subclasses and light chain specificities. *Journal of Pharmaceutical and Biomedical Analysis*, *184*, 113166. Available from https://doi.org/10.1016/j.jpba.2020.113166.

Wang, H., Levi, M. S., Del Grosso, A. V., McCormick, W. M., & Bhattacharyya, L. (2017). An improved size exclusion-HPLC method for molecular size distribution analysis of immunoglobulin G using sodium perchlorate in the eluent. *Journal of Pharmaceutical and Biomedical Analysis*, *138*, 330–343.

Wang, T., Liu, Y. D., Cai, B., Huang, G., & Flynn, G. C. (2015). Investigation of antibody disulfide reduction and re-oxidation and impact to biological activities. *Journal of Pharmaceutical and Biomedical Analysis*, *102*, 519–528.

Wu, G., Yu, C., Wang, W., & Wang, L. (2018). Interlaboratory method validation of icIEF methodology for analysis of monoclonal antibodies. *Electrophoresis*, *39*, 2091–2098. Available from https://doi.org/10.1002/elps.201800118.

Xu, C.-F., Chen, Y., Yi, L., Brantley, T., Stanley, B., Sosic, Z., & Zang, L. (2017). Discovery and characterization of histidine oxidation initiated cross-links in an IgG1 monoclonal antibody. *Analytical Chemistry*, *89*(15), 7915–7923. Available from https://doi.org/10.1021/acs.analchem.7b00860.

Yang, B., Li, W., Zhao, H., Wang, A., Lei, Y., Xie, Q., & Xiong, S. (2019). Discovery and characterization of CHO host cell protease-induced fragmentation of a recombinant monoclonal antibody during production process development. *Journal of Chromatography B*, *1112*, 1–10.

Yang, R., Tang, Y., Zhang, B., Lu, X., Liu, A., & Zhang, Y. T. (2015). High resolution separation of recombinant monoclonal antibodies by size-exclusion ultra-high performance liquid chromatography (SE-UHPLC). *Journal of Pharmaceutical and Biomedical Analysis*, *109*, 52–61. Available from https://doi.org/10.1016/j.jpba.2015.02.032, ISSN 0731-7085.

Yang, J., Wang, S., Liu, J., & Raghani, A. (2007). Determination of tryptophan oxidation of monoclonal antibody by reversed phase high performance liquid chromatography. *Journal of Chromatography A*, *1156*(1–2), 174–182.

Yan, B., Valliere-Douglass, J., Brady, L., Steen, S., Han, M., Pace, D., ... Wang, W. (2007). Analysis of post-translational modifications in recombinant monoclonal antibody IgG1 by reversed-phase liquid chromatography/mass spectrometry. *Journal of Chromatography A*, *1164*(1–2), 153–161.

Yumioka, R., Sato, H., Tomizawa, H., Yamasaki, Y., & Ejima, D. (2010). Mobile phase containing arginine provides more reliable SEC condition for aggregation analysis. *Journal of Pharmaceutical Sciences*, *99*(2), 618–620. Available from https://doi.org/10.1002/jps.21857, ISSN 0022-3549.

Zhang, L., Fei, M., Tian, Y., Li, S., Zhu, X., Wang, L., ... Xie, M. H. (2020). Characterization and elimination of artificial non-covalent light chain dimers in reduced CE-SDS analysis of pertuzumab. *Journal of Pharmaceutical and Biomedical Analysis*, *190*, 113527. Available from https://doi.org/10.1016/j.jpba.2020.113527, ISSN 0731-7085.

Zhang, L., Patapoff, T., Farnan, D., & Zhang, B. (2013). Improving pH gradient cation-exchange chromatography of monoclonal antibodies by controlling ionic strength. *Journal of Chromatography A*, *1272*, 56–64.

Zhang, X., Voronov, S., Mussa, N., & Li, Z. (2017). A novel reagent significantly improved assay robustness in imaged capillary isoelectric focusing. *Analytical Biochemistry*, *521*, 1–7.

Zhang, T., Zhang, J., Hewitt, D., Tran, B., Gao, X., Qiu, Z. J., ... Kao, Y. H. (2012). Identification and characterization of buried unpaired cysteines in a recombinant monoclonal IgG1 antibody. *Analytical Chemistry*, *84*(16), 7112–7123.

Zhou, K., Cao, X., Bautista, J., Chen, Z., Hershey, N., Ludwig, R., ... Das, T. K. (2020). Structure-function assessment and high-throughput quantification of site-specific aspartate isomerization in monoclonal antibody using a novel analytical tool kit. *Journal of Pharmaceutical Sciences*, *109*(1), 422–428.

CHAPTER 6

Higher order structure and free thiol

Yan Chen[1] and Kai Zheng[2]
[1]Pharma Technical Development, Genentech Inc, South San Francisco, CA, United States
[2]Pharmaceutical Development, Genentech Inc, South San Francisco, CA, United States

6.1 Introduction

Monoclonal antibodies (mAbs), which belong to the immunoglobulin family, are the most dominant platform of protein therapeutics. At present, approximately 100 therapeutic mAbs have been approved or in regulatory review in the United States and Europe for the treatment of immunologic diseases, reversal of drug effects, and cancer therapy (Reichert, 2020). It is widely known that mAbs are highly complex macromolecules and characterizing the higher order structure (HOS) of therapeutic mAbs is important for ensuring product efficacy and safety (Wei, Shacter, Schenerman, Dougherty, & McLeod, 2011). Proper folding of the protein is critical for drug efficacy and stability, whereas misfolding may induce loss of biological function and trigger unwanted immune responses (Scheiblhofer, Laimer, Machado, Weiss, & Thalhamer, 2017). Results from the HOS studies are often included in the regulatory filing documents to demonstrate that a given therapeutic mAb shows the expected properly folded structure. In addition, quantifying HOS differences or similarities between two mAb samples is often used to demonstrate comparability in different batches of the same drug or confirming biosimilarity when comparing a biosimilar to the proprietary form (FDA, 2015; Wang, Li, & Davies, 2014).

Protein HOS includes secondary, tertiary, and quaternary structures due to their large size and the rotational characteristics of protein alpha carbons among other factors. There is even quinary structure that corresponds to the spatial and organizational relationship between protein and cell interior (Khan et al., 2017). At approximately 150 kDa, antibodies are composed of repeating protein-folding units, the so-called "immunoglobulin (Ig)" domains (Feige, Hendershot, & Buchner, 2010). The Ig domains are

Monoclonal Antibodies
DOI: https://doi.org/10.1016/B978-0-12-822318-5.00003-X
Copyright © 2021 Elsevier Inc.
All rights reserved.

composed primarily of beta sheet, with only a very small stretch of alpha helix. A single-Ig domain consists of seven or nine beta strands that form a sandwich of two antiparallel sheets, with an intrachain disulfide bridge linking the sheets for extra stability. An antibody consists of 12 of these Ig domains in four chains including two domains per light chain and four domains per heavy chain. The light and heavy chains are covalently linked by interchain disulfide bonds. In addition, there are 12 intrachain disulfide bonds in total per IgG molecule, and this general pattern of intrachain disulfide bridge formation is maintained for each of the IgG subclasses (Correia, 2010). Antibodies contain two fragment antigen-binding (Fab) arms that include one constant and one variable domain of each of the heavy and the light chain. The tail region of an antibody that interacts with cell surface receptors and some proteins of the complement system is called the fragment crystallizable region (Fc region) (Dreker, Schwarz, Reichel, & Hilschmann, 1976; Waxdal, Konigsberg, Henley, & Edelman, 1968). An additional degree of conformational change can be created by the relatively free Fab and Fc domains, and the complexity of the mAb HOS makes accurate determination of its status a challenge.

There are many analytical technologies available for HOS analysis of proteins, and Table 6.1 lists some of the commonly used techniques. Each

Table 6.1 Analytical methods for protein HOS determination.

Protein structure	Structure information	Methods of analysis
Secondary	Local folding patterns, shape of peptide chains (α-helices, β-sheets, etc.)	X-ray crystallography, Cryo-EM, NMR, far UV CD, FTIR, Raman, HDX-MS, and HRF-MS
Tertiary	Overall folding, three-dimensional structure	X-ray crystallography, Cryo-EM, NMR, FTIR, near UV CD, Raman, disulfide analysis, and intrinsic and extrinsic fluorescence
Quaternary	Interaction between polypeptide units, subunit arrangement	X-ray crystallography, Cryo-EM, NMR, light scattering, native MS, and AUC

HOS, higher order structure; *MS*, mass spectrometry; *Cryo-EM*, cryoelectron microscopy; *NMR*, nuclear magnetic resonance; *CD*, circular dichroism; *FTIR*, Fourier-transform infrared; *HDX-MS*, Hydrogen–Deuterium Exchange mass spectrometry; *HRF-MS*, hydroxyl radical footprinting mass spectrometry; *AUC*, analytical ultracentrifugation.

technique is unique with respect to its ability to monitor different aspects of the protein structure. No single technique provides 100% HOS information for a protein in solution, and a careful selection of orthogonal methods is required. A thorough understanding of both the capabilities and limitations of each technique is critical for determining the most appropriate strategy. Table 6.2 summarizes some of the considerations including method resolution related to HOS understanding and ease of analysis.

X-ray crystallography can provide the complete three-dimensional structure of a protein at high resolution (Smyth & Martin, 2000). However, this technique requires the molecule to form a well-ordered crystal, which is not possible for all proteins. In addition, since all structures are determined from the solid state of the protein, it is difficult to derive solution-state dynamic structural information with this technique. Due to the natural flexibility of antibodies, it is difficult to determine the structure of a full-length antibody using X-ray crystallography, and only a few have been reported (Harris, Larson, Hasel, & McPherson, 1997; Harris, Skaletsky, & McPherson, 1998; Saphire et al., 2001; Scapin et al., 2015). At present, the Protein Data Bank (Berman, Bourne, Westbrook, & Zardecki, 2003) contains over 3500 antibody structures. However, most of them are antibody fragments such as Fabs, Fvs, scFvs, and Fcs. Over the last several years, cryoelectron microscopy (cryo-EM) has emerged as a viable tool for structural elucidation of large molecules and molecule complexes (usually >100 kDa) at atomic resolution without the bottleneck of crystallization

Table 6.2 Comparison of popular HOS analytical methods.

Method	Resolution	Difficulty/challenges
X-ray crystallography	High	High
Cryo-EM	High	Medium
High-field NMR	High	High
MS (HDX-MS, HRF-MS, or Native MS)	Medium	Medium
FTIR, Raman	Medium	Medium
CD	Medium	Low
Fluorescence	Low	Low
Light scattering	Low	Low
Analytical ultracentrifugation	Low	Low

HOS, higher order structure; *MS*, mass spectrometry; *Cryo-EM*, cryoelectron microscopy; *NMR*, nuclear magnetic resonance; *CD*, circular dichroism; *FTIR*, Fourier-transform infrared; *HDX-MS*, Hydrogen–Deuterium Exchange mass spectrometry; *HRF-MS*, hydroxyl radical footprinting mass spectrometry.

(Cressey & Callaway, 2017; Wang & Wang, 2017). In cryo-EM, buffer-based samples are rapidly frozen with a bath of liquid ethane cooled by liquid nitrogen. As a result, the molecules of water with supporting carbon films cannot form a crystalline lattice before being frozen into the amorphous form of an ice crystal, allowing the biomolecules in the sample to be retained in their native conformation. Another advantage of cryo-EM, as compared to X-ray crystallography, is that it allows for thousands of flash-frozen images of proteins (Li et al., 2013). One can then reconstruct these snap-shots into movie-like scenes to study the protein dynamic structural behavior. There has been a huge surge of interest in cryo-EM within the structural biology community, and it is anticipated more data on protein analysis will become available in the near future.

High-field nuclear magnetic resonance (NMR) is another established and popular technique that provides high-resolution structural information for proteins (Bax, 1989). Historically, the interpretation of the NMR spectra of large proteins has been challenging due to difficulty in resolving overlapping peaks, which often become broader and weaker with increasing molecular weights (MWs). Therefore early application was limited to smaller proteins (Fernández & Wider, 2003). Recent technical advancements in both hardware and software of NMR have improved the signal-to-noise (S/N) ratio and resolution for large proteins such as mAbs (Chen, Freedberg, & Keire, 2015; Franks et al., 2016; Japelj et al., 2016; Kheddo, Cliff, Uddin, van der Walle, & Golovanov, 2016; Poppe et al., 2013). There has been an increasing interest in the biopharmaceutical community to apply NMR to obtain atomic-level fingerprint of the protein structure in comparability assessment and biosimilarity evaluations. Detailed discussion on this approach is provided in this review.

In general, there are a number of disadvantages associated with the high-resolution techniques described above, which prevent them from routine adoption for HOS analysis of protein therapeutics. They often require relatively large amounts of sample and laborious sample pretreatment and/or derivatization procedure. The equipment can be costly, and a high level of expertise is needed for conducting the studies, performing data analysis, and interpreting results. These limitations have led to the development of alternative methods that provide structural information with low-to-moderate resolution but are more user friendly. The most commonly used methods include spectroscopic, thermodynamic, and hydrodynamic techniques as well as indirect techniques that measure structure-related thermal stability or biological function. Examples include mass spectrometry (MS)-based methods, circular dichroism (CD) (Greenfield, 2006), Raman (Rygula et al., 2013), or Fourier-transform infrared (FTIR) spectroscopies (Lin, Glover, & Sreedhara, 2015), differential scanning calorimetry (DSC) (Johnson, 2013), intrinsic and extrinsic fluorescence (Basak & Chattopadhyay, 2014), dynamic light scattering (DLS) (Murphy, 1997), and sedimentation velocity analytical ultracentrifugation (SV-AUC) (Gabrielson et al., 2007).

MS has shown to be a powerful tool for the HOS characterization of therapeutic proteins. For example, native MS analyzes intact proteins and their complexes without destroying noncovalent interactions, which is critical to HOS analysis (Haberger et al., 2016). Ion mobility MS under native-like conditions can provide information on the stoichiometry, composition, connectivity, topology, and conformational heterogeneity of protein complexes, such as composition of multimers or noncovalent protein–protein and protein–ligand complexes (Ferguson & Gucinski-Ruth, 2016). In addition, advanced techniques of chemical modification and proteolysis allow the study of protein conformational changes or protein–protein interactions in protein complexes using MS analysis. Commonly used approaches include Hydrogen–Deuterium Exchange (Englander, 2006), Hydroxyl Radical Footprinting (Garcia, Deperalta, & Wecksler, 2019), cross-linking chemistry (Singh, Panchaud, & Goodlett, 2010), and limited proteolysis (Gao et al., 2005). However, similar to the high-resolution techniques, MS is expensive, low throughput, time-consuming, and has poor reproducibility in some cases.

Spectroscopic methods, such as FTIR (Lin et al., 2015), Raman (Rygula et al., 2013), or CD (Greenfield, 2006), can characterize secondary and tertiary structures of proteins in a more user friendly way. They are classical biophysical tools in the HOS determination of therapeutic mAbs due to numerous advantages such as modest sample requirements, relatively uncomplicated data acquisition and analysis, and reasonable instrument cost. However, the large size of mAbs causes these methods to be less sensitive to chemical or physical structural heterogeneity. For example, they are unlikely to detect subtle tertiary conformational changes that may be caused by small differences in the formulation environment (Rodríguez-Mendieta, 2015).

As with other spectroscopic methods, intrinsic protein fluorescence can be used to compare different batches or to monitor denaturation. Aromatic amino acids, chiefly tryptophan, absorbs UV light (280–290 nm) and emits at a longer wavelength (ca. 340 nm) (Basak & Chattopadhyay, 2014). The hydrophobic tryptophan would typically be shielded from the aqueous environment. Therefore changes in fluorescence intensity and emission wavelength between different protein samples correlate directly with perturbation of proteins' tertiary structure.

Other popular biophysical techniques such as DLS and SV-AUC provide some type of structural information at low resolution but is far from defining complete information about HOS. For example, DLS only provides an assessment of aggregation but no specific secondary structure information can be derived. Only very indirect tertiary structure data can be discerned from the experimentally determined hydrodynamic radius (Murphy, 1997). AUC provides the MW information of protein mixtures with different size ranges, using a series of standards separated together with the target biopharmaceutical in a high-density medium, at high rates of spinning in a commercial centrifugation instrument. The approach allows for the accurate and precise determination of monomer, aggregates, and fragments of the protein studied. However, it does not define the true conformation or shape of the protein under any biological solution conditions (Gabrielson et al., 2007).

For protein structural understanding, disulfide bridge mapping and free thiol analysis are often performed. Disulfide bonds are covalent linkages between the sulfur-containing side chains of cysteines and contribute to protein folding and appropriate tertiary structure. In theory, all cysteine residues in standard IgG antibodies are in the disulfide-bonded state. However, unpaired cysteine residues, often referred to as free thiols or free sulfhydryls, have been routinely detected in all subclasses of IgG antibodies, including recombinant and human IgG molecules (Gevondyan, Volynskaia, & Gevondyan, 2006; Huh, White, Brych, Franey, & Matsumura, 2013; Lacy, Baker, & Brigham-Burke, 2008; Schauenstein, Dachs, Reiter, Gombotz, & List, 1986). Various analytical methods are available to detect and quantify free thiols in a protein. Total free thiol content in a native or denatured protein can be readily measured by reacting with Ellman's reagent, DTNB [5,5-dithio-bis-(2-nitrobenzoic acid)], or fluorescent tags (Ellman, 1959; Zhang & Czupryn, 2002). In the meantime, site-specific (or domain-specific) detection and quantification of free thiol is often obtained by LC-MS analysis (Cheng et al., 2017; Xiang, Chumsae, & Liu, 2009).

In this chapter, we first review recent advances in the NMR HOS profiling. Second, traditional standard technologies (FTIR, Raman, and fluorescence) for protein HOS characterization are discussed. Lastly, tools for "free thiol" determination are presented. Although the applications of these analytical techniques are much broader than HOS, given the scope of this chapter, we will focus on HOS characterization of the methods mentioned above.

6.2 Nuclear magnetic resonance

NMR is an analytical chemistry technique first demonstrated in 1946 by Felix Bloch and Edward Mills Purcell (Bloch, 1946). When molecules are placed in a strong external magnetic field and broadband radio frequency pulses are applied, certain atomic nuclei that are intrinsically magnetic will begin to resonate. Isotopes, such as ^{1}H, ^{13}C, ^{15}N, and ^{31}P, have a magnetic moment, whereas other isotopes such as ^{12}C or ^{16}O are NMR inactive. The electronic molecule orbitals in the chemical surroundings of the magnetic nucleus create a small local magnetic field that opposes the applied field, causing the NMR signal frequency to shift. The magnitude of the shielding depends on the strength of the external magnetic field, the type of nucleus, the chemical structure, and the structural geometry of the molecule. Consequently, different atoms within a molecule resonate at different frequencies. Study of the frequency, linewidth, and coupling patterns provides detailed information about the structure, dynamics, reaction state, and chemical environment of the molecule studied.

A resonance spectrum for a molecule can be obtained by varying the magnetic field at a constant frequency of electromagnetic radiation or by keeping the magnetic field constant and varying electromagnetic radiation. The resonant frequencies of the nuclei are measured and displayed as peaks in a NMR spectrum. The *x*-axis corresponds to *chemical shift*, the resonant frequency of a nucleus relative to a reference standard. It is normally measured in Hz and expressed in fractional units δ (parts per million, or ppm). The area of each peak (*y*-axis) represents the number of nuclei that resonates at each specific frequency magnitude and is proportional to the molar concentration of the sample. Both the sensitivity and the resolution increases with the magnetic field strength of the superconducting magnet. In the case of Fourier-transform instruments, the S/N ratio also depends on the acquisition time (Farrar & Becker, 2012), although it is not a major factor in real practice in optimizing the sensitivity.

NMR has been applied at various stages of the drug development process for both small molecule (Williamson & Marquez, 2017) and protein therapeutics (Kiss, Fizil, & Szántay, 2018). During early drug discovery, it is commonly used for the structure elucidation and conformational analysis of candidate molecules (Holzgrabe, 2010). It is also one of the most important tools for characterizing protein–ligand complexes to identify hits from compound library screening (Pellecchia, Sem, & Wüthrich, 2002). In the development phase for small molecule therapeutics, structure elucidation and quantitation tasks are often associated with characterizing impurities and degradation products from the lead molecules (Maggio, Calvo, Vignaduzzo, & Kaufman, 2014). In manufacturing, NMR applications in quality control (QC) and troubleshooting are currently limited due to various reasons including complex data analysis and less competitive instrument price, etc.

NMR plays a major role in the determination of the structures and dynamics of proteins and other biological macromolecules (Bax, 1989). It allows direct determination of three-dimensional structures of protein molecules at the atomic level in the solution phase and is a very useful method to study reaction kinetics. This is a clear advantage over X-ray crystallography, another high-resolution technique but requires crystallization of proteins. To obtain the exact three-dimensional structural information of proteins with NMR, it is critical to ensure that the signals in the spectrum are assigned to the correct nuclei. This requires a series of experiments, each of which provides partial clues about the nature of the atoms in the protein molecule. However, structure determination for proteins with NMR spectroscopy does have its limitations. It is relatively easy to solve the structure for proteins with MW below 30 kDa using routine NMR techniques (Paterson, Englander, & Roder, 1990). Although there has been some success reported for proteins up to 40–50 kDa (Rovnyak, Hoch, Stern, & Wagner, 2004), resonance assignment and de novo structure determination for large proteins such as mAbs are significantly complicated and practically unattainable due to various reasons. The number of signals increases with MW; this complexity increases the likelihood of resonance overlap, making it difficult to obtain site-specific information. In addition, larger proteins typically have faster relaxation rates, which results in increased linewidths of the NMR peaks that tend to merge with one another (Foster, McElroy, & Amero, 2007).

Even in the absence of sequence-specific resonance assignments to enable detailed and accurate structure determination for mAbs, NMR is unique in its ability to provide highly detailed fingerprint-like information about the protein HOS at atomic level (Arbogast, Delaglio, Schiel, &

Marino, 2017; Poppe et al., 2013). This HOS profiling capability of NMR offers valuable structural and functional insights into therapeutic mAb molecules under investigation. The chemical shift depends on amino acid type, secondary, tertiary, quaternary, and quinary structures of the protein. The spectral line shape is affected by quaternary and quinary structures and exchange kinetics between different forms (Chen et al., 2016). Such information may be utilized to study structural changes induced by experimental conditions such as changes in temperature, pH, ionic strength, excipients, and residue mutations (Aubin, Hodgson, Thach, Gingras, & Sauvé, 2015). It has been used to predict protein–protein and protein–excipient interactions, as well as the propensity of proteins to aggregate (Pandya, Howard, Zloh, & Dalby, 2018). In recent years, it has also become an excellent tool to differentiate samples of interest and assess product comparability (Chen et al., 2016) and/or biosimilarity at the HOS level (Japelj et al., 2016).

Both one- or two-dimensional (1D or 2D) NMR methods have been used for the HOS profiling of mAbs. One-dimensional NMR method is especially useful for a quick assessment of protein HOS in various formulation conditions and offers easy distinction between different materials. A number of 1D ^{1}H NMR profiling protocols have been applied for HOS analysis of therapeutic mAbs (Franks et al., 2016; Poppe et al., 2013). However, direct analysis of mAbs in aqueous condition with 1D NMR is challenging due to spectral overlap with much stronger signals from the formulation buffer. Significant portions of the spectrum cannot be used for protein analysis. A complete buffer exchange is often performed for protein samples to be compared prior to NMR analysis. For example, Chen et al. (2016) used purified IgG1 fractions taken at the highest UV absorbance from the SEC-FPLC (fast protein size-exclusion liquid chromatography) elution in the 1D ^{1}H NMR analysis to remove buffer interference and increase detection sensitivity. Other efforts to suppress the excipient signals include the development of diffusion filtering methods that are based on different diffusion properties of the small excipient molecules and the large proteins. They were shown to be effective in reducing sucrose (Poppe et al., 2013) and trehalose peaks (Franks et al., 2016) in the 1D ^{1}H NMR spectra of mAbs.

One-dimensional ^{1}H NMR (Franks et al., 2016) was successfully applied to the detection of partial pH denaturation of an IgG4 mAb known to undergo denaturation at ~pH 3.0. Changes in line shape and intensities were observed when pH was decreased from 6.5 to 4.5. Detailed analysis of spectral components including sharper and broader HOS components and a less ordered component indicated selective unfolding of protein segments on decreasing pH. Results suggest that the 1D ^{1}H NMR method developed is sensitive to detect subtle structural changes well before the mAb is fully denatured. In another study, 1D ^{1}H NMR spectra were collected to compare the HOS of rituximab and infliximab, two IgG1 molecules that are homologous in sequence, superimposable in Fab crystal structure and identical in Fc structure (Chen et al., 2016). Different NMR spectral patterns were observed comparing the two intact IgG1 molecules even though both showed broad resonance peaks. To achieve better spectral quality, in situ digestion (i.e., directly in the NMR tube) was performed using IdeS (protease, added at an amount that does not contribute to the NMR spectra) to release the Fc fragment ($\sim$50 kDa) from F(ab′)2 ($\sim$100 kDa). Upon the IdeS digestion, the resonances narrowed in linewidth and the S/N ratio improved, at 1.5- and 2.5-folds for rituximab and infliximab, respectively. Results from both the intact and in situ digested samples indicated the two IgG1 molecules are significantly different in quinary structure (complexes in which each composing unit maintains native quaternary structure) of the stable noncovalent dimer and oligomers of the mAbs. In addition, minor differences in the secondary structure were observed in the two antibodies studied. Other HOS techniques were also applied, and all data collectively showed rituximab was mostly monomeric while infliximab was in mono-oligomer equilibrium driven by its Fab portion of the molecule. The study result confirmed the advantage of NMR as a HOS technique over X-ray crystallography, which has limitations in solution-state quaternary and quinary structure comparisons. In the same study, superimposable NMR spectra were observed visually when different lots from the same IgG1 molecule were compared, indicating that their HOS was consistent from lot-to-lot.

When used alone, 1D NMR for HOS profiling of mAbs may still be challenging. There are many distinct nuclei present in a single large mAb molecule and significant overlap in the chemical shifts is expected. Multidimensional NMR is desirable because it offers better capability in resolving peaks of excipients and impurities as well as differentiating similar protein therapeutics. Two-dimensional heteronuclear $^{1}H-^{15}N$ and $^{1}H-^{13}C$ correlation spectra have been applied to characterize HOS profiles of mAbs (Chen et al., 2015). They are superior to homonuclear two-dimensional spectra due to increased chemical shift dispersion with separation of chemical shift into a second dimension (^{13}C or ^{15}N), less homonuclear J-splitting, and concomitant reduced peak overlap. In a 2D

NMR spectrum, resonances are represented as cross-peaks that carry information about various through-bond or through-space atomic connections within a molecule. They are less crowded compared to 1D NMR spectra. For example, a 2D $^1H - {}^{15}N$ heteronuclear single-quantum correlation (HSQC) spectrum only shows cross-peaks that represent the H-N units in the protein molecule when the excipients in the sample do not contain NH or NH_2 groups.

Initial efforts to generate 2D NMR spectral maps of mAbs often involved artificial introduction of the ^{13}C or ^{15}N isotopes into the proteins due to the low natural abundance of these nuclei. Techniques were developed to produce isotope-labeled proteins using Chinese hamster ovaries (Kato, Yamaguchi, & Arata, 2010) and *E. coli* cell expression platforms (Reddy et al., 2018). For example, a triply labeled human IgG Fc fragment containing ^2H, ^{13}C, and ^{15}N was expressed using an *E. coli*-based system and analyzed in a forced degradation study (Liu et al., 2008). Two-dimensional ^1H-^{15}N HSQC NMR study yielded nearly complete resonance assignments and provided great details about the structural impact of methionine oxidation on the Fc region.

However, isotopic labeling is not typically available when analyzing the formulated drug products of therapeutic antibodies. In recent years, acquisition of 2D NMR spectra at natural isotopic abundance has become a practical approach to study protein therapeutics due to several technical advancements. High-field NMR with increased spectrometer field strength ranging from 300 to over 1000 MHz has become more accessible. In addition, the introduction of cryogenic probes has dramatically increased the inherent sensitivity of NMR experiments (Yao & Yoon, 2000). In this case, the $^1H - {}^{13}C$ methyl fingerprint approach offers more advantages than the $^1H - {}^{15}N$ amide-based approach. There is greater natural abundance of ^{13}C relative to ^{15}N (1.1% vs 0.37%). The amino acids-containing methyl groups in their side chains are good reporters of proper folding of a protein as they are sensitive to the hydrophobic packing of the protein. Meanwhile, the resonances of the methyl groups are most often not involved in chemical exchange processes, which make them less sensitive to study conditions (Arbogast, Brinson, & Marino, 2015). Furthermore, the ^1H and ^{13}C methyl resonances are in a distinct region of the spectrum and are usually well separated from the signals of the excipients. Lastly, the freely rotating methyl moieties provide slower relaxation rate, yielding sharper lines and greater sensitivity when compared to amide groups (Arbogast et al., 2015).

Arbogast et al. (2015) demonstrated the feasibility of rapid fingerprinting of the methyl resonances for structural mapping of an intact mAb at ^{13}C natural isotopic abundance using 2D NMR. They also showed that the 2D $^{1}H-{}^{13}C$ correlation NMR spectra of papain-digested Fc and Fab fragments can provide accurate domain structural information and the HOS of the fragments was retained when compared to the intact mAb molecule. The Fab and Fc fingerprint spectra can be rapidly acquired in as short as approximately 30 min with fast-pulsed sequences and nonuniform sampling techniques. The same group (Arbogast et al., 2017) also showed 2D NMR can detect even the most subtle changes on HOS resulting from engineered NISTmAb glycoform populations.

For molecules as large as a 150 kDa mAb, both the 2D heteronuclear NMR and most of the 1D diffusion-filtered ^{1}H NMR often involve sophisticated data acquisition and processing. For HOS profiling, information related to changes in specifically identified peak positions, peak heights and linewidths is often analyzed. Visual inspection of the NMR fingerprints may be the simplest way for structural similarity assessment (Visser et al., 2013). However, it only provides qualitative assessment and can vary from one analyst to another. Manual spectral analysis of complex protein NMR spectra is challenging for analysts, and this problem is further compounded when dealing with large datasets, where many samples are compared. The solution to this is to use multivariate analysis, a powerful tool that can transform complex information into more useful sets of information, and that can extract vital differences from data that may look similar by other methods. Multivariate principal component analysis (PCA) (Arbogast et al., 2017; Japelj et al., 2016) has been successfully applied in the direct analysis of the matrix of spectral data, correlating spectra according to similarities and differences in their overall shapes, rather than according to parameters of individually identified peaks. This is particularly a powerful tool for analyzing NMR spectra of mAbs, where some of the individual peaks might not be well resolved.

Brinson and coworkers (Brinson et al., 2019) at the National Institute of Standards and Technology (NIST) recently completed a large international round robin study that demonstrates 2D NMR as a reliable way to compare the 3D structures of therapeutic mAbs. Twenty-six industrial, government, and academic laboratories worldwide participated in this study performing 2D NMR analysis on the NISTmAb, a publicly available reference material. Both $^{1}H-{}^{15}N$ and $^{1}H-{}^{13}C$ NMR spectra were acquired on 39 different NMR spectrometers with magnetic fields ranging from 500 to 900 MHz, of different vintages of vendor instrumentation and equipped with conventional or high-sensitivity NMR probes. Chemometric analyses using PCA from over 400 spectral maps showed that a simple 2D NMR study can produce a spectral map that offers a comprehensive, atomic-level fingerprint of the primary, secondary, tertiary, and quaternary structure(s) of mAb-based biotherapeutics. It is noted that most of the successful stories of the 1D and 2D NMR fingerprinting studies were published in the past 5 years. With further technical advancement, this technology will certainly find wide applicability as a powerful tool for structural assessment of protein therapeutics including mAbs.

6.3 Fourier transform infrared spectroscopy and Raman spectroscopy

6.3.1 Fourier transform infrared spectroscopy

FTIR spectroscopy is an absorption spectroscopy, which measures the vibrational frequency of covalent bonds. When IR radiation hits a sample, part of the radiation is absorbed and the remaining radiation is recorded on a detector, which builds a frequency interferogram. Because the interferogram records the intensity of each individual frequency, it cannot be analyzed directly. Therefore Fourier transform is used to mathematically decode the frequencies and transform the interferogram to a FTIR spectrum for further interpretation (Cooley & Tukey, 1965). Typically, an IR spectroscopy covers a broad spectrum range, for example, wavelengths from 2.5 to 15.4 μm (Deshpande, 2002). With the Fourier transform, the length dimension in micrometer is converted to the reciprocal length dimension in wavenumbers (cm^{-1}). This unit is proportional to wave frequency.

A FTIR spectrometer includes a light source to emit infrared radiation, an interferometer (typically consists of a beam splitter, fixed and movable mirrors) to encode the spectra, a sample compartment that provides different techniques for infrared radiation adsorption by the sample, a detector to measure the interferogram of the infrared radiation after passing the sample compartment, and a computer with a specific software to process the interferogram using Fourier transform to a FTIR spectrum for data interpretation. The FTIR spectrometer offers many benefits for sample analysis as it is a nondestructive method with a precise measurement (no external calibration required). The FTIR measurement is also relatively quick (e.g., less than a second for a single scan) with a high-spectral resolution and a high S/N ratio.

At the molecular level, covalent bonds behave like springs and exhibit different vibrational modes, for example, stretching vibration, bending vibration, and twisting vibration (Griffiths & De Haseth, 2007; Pavia, Lampman, Kriz, & Vyvyan, 2008; Smith, 2011). These vibrations only happen at certain allowable frequencies. When infrared radiation hits the sample, the radiation with the frequency matches with that of a vibrational mode will be absorbed by the covalent bonds to increase its vibration, that is, turns the covalent bonds from the ground state to an excited state. For an organic molecule, the FTIR spectrum represents a combined result of absorption and transmission from all allowable vibrational modes of its covalent bonds, which can serve as a fingerprint of the molecule. Therefore FTIR has been widely used for material analysis in organic chemistry and pharmaceutical industry and signature frequencies of organic molecule functional groups were well characterized (Lin-Vien, Colthup, Fateley, & Grasselli, 1991).

FTIR is also a well-established and convenient method for HOS characterization for proteins. FTIR can measure proteins in either solution or solid state, and most of the time it does not require additional sample preparation steps. A typical protein spectrum can contain up to nine characteristic infrared absorption regions—amide A, B, I, II, ..., VII bands (Table 6.3) (Kong & Yu, 2007; Krimm & Bandekar, 1986). Among these bands, there are two distinct and structural sensitive bands: amide I band (1700–1600 cm^{-1}), which is mainly due to the absorption of the C = O stretching vibration of the backbone peptide bonds, and amide II band, which is mainly due to the absorption of the N–H bending and C–N stretching vibrations of the backbone peptide bonds (Dousseau & Pezolet,

Table 6.3 Frequency ranges of amide bands.

Amide bands	Approximate frequency (cm^{-1})	Vibrational mode
Amide A	3300	NH stretching
Amide B	3100	NH stretching
Amide I	1600–1690	C = O stretching
Amide II	1480–1575	CN stretching, NH bending
Amide III	1229–1301	CN stretching, NH bending
Amide IV	625–767	OCN bending
Amide V	640–800	Out-of-plane NH bending
Amide VI	537–606	Out-of-plane C = O bending
Amide VII	200	Skeletal torsion

This table is derived from Table 1 in Kong and Yu (2007).

1990; Haris & Severcan, 1999; Miller, Bourassa, & Smith, 2013). The amide I band is widely used for HOS assessment as it is most characteristic and sensitive to protein secondary structures and structural stability (Bath & Zscherp, 2002; Byler & Susi, 1986; Chen, Hsieh, Cheng, & Liao, 2001; Goormaghtigh, Ruysschaert, & Raussens, 2006; Li, Lantz, & Du, 2019; Miller et al., 2013; Susi & Byler, 1986).

One challenge to use the amide I band for structural analysis is that it is a broad peak with low resolution of individual structures. This is because protein secondary structures (α-helix, β-sheet, β-turn, and random coil) possess different molecular geometries and H-bonding patterns, which lead to different absorption peaks and have a substantial overlap in the amide I region (Khattari, 2017; Kong & Yu, 2007; Miller et al., 2013). Therefore the second derivative and curve fitting methods are often used to deconvolute the complex amide I band and allow quantitative assessment of protein secondary structures (Baldassarre, Li, Eremina, Goormaghtigh, & Barth, 2015; Dong, Huang, & Caughey, 1990; Haris & Severcan, 1999; Usoltsev, Sitnikova, Kajava, & Uspenskaya, 2019). The second derivative of a FTIR spectrum reflects the rate of the slope change, which can resolve the overlapping peaks. Peak assignments to individual secondary structures are listed in Table 6.4.

The other challenge for FTIR analysis of protein structures is the interference of water. As proteins are often formulated in aqueous

Table 6.4 Peak assignments to secondary structures in amide I band in H_2O.

Approximate frequency (cm^{-1})	Structure assignment
1623–1625	β-Sheet
1625–1629	β-Sheet
1631–1635	β-Sheet
1636–1640	Antiparallel β-sheet
1641–1643	β-Sheet
1646–1650	Random coil
1654–1658	α-Helix
1660–1666	3_{10}-Helix
1666–1668	β-Turn
1674–1676	β-Turn
1678–1682	β-Turn
1683–1687	β-Turn
1689–1693	Antiparallel β-sheet
1694–1698	β-Sheet

This table is derived from Table 2 in Kong and Yu (2007).

solutions, H—O—H bending vibration of the water molecule contributes a strong absorption at $\sim 1645\ cm^{-1}$, which overlaps with the amide I band (Yang, Yang, Kong, Dong, & Yu, 2015). Therefore the water absorption peak has to be accurately subtracted prior to structural analysis of protein infrared peaks. The water subtraction in FTIR analysis has been well studied and used in many reports (Dong, Huang, & Caughey, 1992; Kal'nin & Ven'yaminov, 1988; Lee, Haris, Chapman, & Mitchell, 1990; Surewicz, Szabo, & Mantsch, 1987; Susi, Timasheff, & Stevens, 1967). In general, researchers can run both the sample solution and the reference solution (same solution composition without protein) under the same conditions, for example, sample compartment pathlength, temperature, number of scans, and then subtract the reference spectrum from the sample spectrum. Besides the water absorption from the aqueous phase, water vapor as well as carbon dioxide in atmosphere can also contaminate the sample spectrum. Typically, the optical path and the sample compartment are purged with nitrogen gas during the analysis to reduce the atmospheric impact. In addition, prestored reference spectra of water vapor and carbon dioxide can be subtracted from the sample spectra to further compensate the atmospheric impact.

During development of therapeutic mAbs, FTIR, is often used to collect HOS data as a quick and convenient method. It is well accepted by regulatory agencies for both clinical and commercial filings. For therapeutic indications that favor a high protein concentration (e.g., 100 mg/mL or above) to reduce dose volume or dose frequency, especially for liquid formulations, the advantages of FTIR are more prominent. For example, many analytical methods require diluting the sample to a low concentration (e.g., <1 mg/mL) to avoid saturation of the signal. FTIR allows direct measurement of the sample in its original formulation without any dilution step, which can reveal its true structural information at high protein concentrations. In addition, protein aggregation is always a concern due to potential impacts on immunogenicity and efficacy. FTIR can be used to assess changes of aggregates under different storage conditions or buffer/excipient compositions without the concern of causing the dissociation of reversible aggregates upon sample dilution.

To qualify the FTIR method as a suitable assay for quantitative assessment of protein HOS, Jiang et al. used a quantitative function (OMNIC QC Compare) to evaluate assay precision and sensitivity (Jiang et al., 2011). A multisite, multiinstrument, and multianalyst study was conducted to evaluate the precision of the FTIR method. Furthermore, Jiang et al.

compared spectra of native, partially unfolded, and fully unfolded proteins, including blending different types of proteins to assess the sensitivity of the FTIR method. A variety of proteins (including IgG1 mAb, IgG2 mAb, Fc conjugates, growth factor, and cytokine) were tested and results showed that the FTIR method has suitable precision. The spectral similarity is greater than 90%, and the standard deviation is no more than 1.1 cm^{-1}. In addition, the FTIR method is sensitive to assess protein secondary structural changes induced by pH and denaturant stresses. The sensitivity depends on the nature and the level of the changes in protein structure and the extent of the corresponding spectral changes in FTIR. For example, when the authors mixed folded species of Protein 1 and Protein 7 with unfolded species, using computer simulation, they observed unfolding through spectral similarity calculations when there is 15% unfolded species in the mixture.

On the other hand, another report by Lin et al. showed that FTIR is not sensitive to detect changes for a mAb after exposed to various stress conditions (Lin et al., 2015). In this study, five stress conditions (low pH, high pH, light, thermal, and oxidation) in a forced degradation study of a mAb were investigated. Both chromatographic assays such as size-exclusion chromatography (SEC) and ion-exchange chromatography (IEC) and HOS characterization techniques including FTIR and CD were performed to compare product quality of the samples side-by-side. The results showed that both FTIR and CD could not identify major changes for the degraded samples, whereas SEC and IEC could detect measurable changes for certain degraded samples (e.g., ~ 7% loss of SEC monomer for the low pH stressed samples and ~32% loss of IEC main peak for the high pH stressed samples). It is noted that there are some major differences between the two studies: Jiang et al. included five different types of proteins, whereas samples from only one mAb were tested in the other study; a quantitative OMNIC QC compare algorithm was used to compare the FTIR spectra in Jiang's study, whereas Lin et al. did not take such a data analysis approach. However, it is clear that FTIR should be used with caution based on what Lin et al. reported. It may not be suitable for a comparability assessment of proteins if the degradation does not cause an extensive change in protein conformation and does not alter the corresponding spectrum substantially.

In summary, FTIR is a useful tool to characterize protein HOS in the biopharmaceutical industry. The FTIR method accepts a wide range of protein concentrations and generates reproducible results rapidly. With the advancement of the FTIR instruments and software, data processing and interpretation can be more straightforward and less labor-intensive. However, researchers also need to be aware of the limitations of FTIR. This method generally assesses the overall protein conformation and may not be sensitive enough for detecting minor structural changes under certain stress conditions. Finally, due to the extensive overlap of the characteristic peaks in the amide I region, some FTIR data may be ambiguous and require an orthogonal method to confirm the interpretation.

6.3.2 Raman spectroscopy

Raman spectroscopy is a scattering spectroscopy or vibrational spectroscopy based on the Raman effect, discovered by Sir C. V. Raman in 1928 (Krishnan & Shankar, 1981). When monochromatic light hits a molecule in vibrational motion, the photons are annihilated to excite the molecule to a virtual state and then are regenerated when the molecule returns to its vibrational state, that is, scattered light is emitted. Most of the new photons are scattered elastically (Rayleigh scattering), and a very small portion (~ 1 in 10^7 photons) spreads inelastically (Raman scattering). The Rayleigh scattering light has the same wavelength as the excitation light; therefore it does not contain molecule specific information. On the other hand, the much weaker Raman scattering light possesses a different wavelength than the excitation light, which corresponds to the vibrational modes of the functional groups in a molecule. The Raman scattered light with a longer wavelength is called Stokes scattering, and the light with a shorter wavelength is called anti-Stokes scattering (Fig. 6.1) (Chalmers & Griffiths, 2002; McCreery, 2005; Pelletier, 2003). In general, intensity of Stokes scattering is much higher than anti-Stokes scattering as the molecular bonds are more favored in the ground state. Therefore the Raman spectrum of a molecule is contributed mostly from the Stokes scattering.

A Raman spectrometer includes a light source (a laser beam) to emit a single-wavelength radiation in the near UV, visible, or near infrared region, a sampling system to pass the monochromatic light to the sample, a system to separate the Raman scattering from Rayleigh scattering, an interferometer to exploit the interference phenomenon, and a charge-coupled device detector. A typical Raman spectrum shows the intensity of Raman scattering against the shift of the monochromatic light frequency. Although theoretically the Raman shift does not depend on the wavelength of the excitation light, the intensity of the Raman signal decreases proportionally to the fourth power of the excitation wavelength

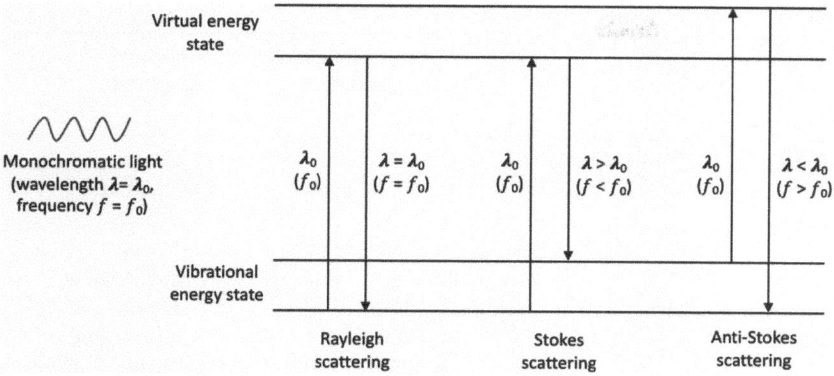

Figure 6.1 Illustration of Raman light scattering.

per the Rayleigh light scattering law. Therefore an appropriate wavelength of the light source needs to be selected with considerations of Raman signal intensity, potential fluorescence emission, and photostability of the sample.

Raman spectroscopy is a nondestructive method and has been widely used in organic chemistry and pharmaceutical industry. The pattern of a Raman spectrum is characteristic of chemical bonds and thus is a fingerprint of the molecule. With the advancement of the Raman technology, Raman spectroscopy can also be used for protein HOS analysis (Krimm & Bandekar, 1986; Rygula et al., 2013). There are numerous characteristic regions in Raman spectra sensitive to protein structures: the amide I region ($C=O$ stretching vibration of the backbone peptide bonds), the amide III region ($N-H$ and $C-H$ bending vibration of the backbone peptide bonds), and the backbone skeletal stretch region ($C_\alpha-C$, $C_\alpha-C_\beta$, $C_\alpha-N$ stretching vibration) (Table 6.5) (Carey, 1982; Ettah & Ashton, 2018; Krimm & Bandekar, 1986; Spiro & Gaber, 1977; Tu & Tu, 1982; Wen, 2007). Similar to FTIR spectra, deconvolution and curve fitting can be used for data analysis of the amide I band to estimate specific secondary structures (Williams, 1986). Besides providing protein backbone information related to the secondary structure, Raman spectroscopy can also acquire amino acid side-chain and disulfide signals, which can reveal protein local environments and the tertiary structure. For example, UV resonance Raman spectroscopy equipped with a deep UV laser (e.g., 244 nm, 257 nm) can measure the spectra of Trp and Tyr chromophore (Asher et al., 1993; Asher, Ludwig, & Johnson, 1986; Cho & Asher, 1993; Wen & Thomas, 1998).

Table 6.5 Peak assignments to secondary structures.

Characteristic region	Approximate frequency (cm^{-1})	Structure assignment
Amide I (~1630–1700 cm^{-1})	1680	β-Turn
	1670–1680	β-Sheet and β-barrel
	1650–1655	α-Helix
	1640	Loose β-sheet
Amide III (~1230–1340 cm^{-1})	1300–1340	α-Helix
	1260	Random coil
	1235–1250	β-Sheet
Skeletal stretch (~870–1150 cm^{-1})	930–950	α-Helix

This table is derived from Table 1 in Wen (2007) and Ettah and Ashton (2018).

In a recent study, Bueno et al. analyzed secondary structures of an IgG1 mAb, rituximab, using deep-ultraviolet resonance Raman spectroscopy (Bueno, Long, Kauffman, & Arzhantsev, 2015). In the initial study, two more biotherapeutic samples besides rituximab were tested to cover a wide range of MWs (4–143 kDa) and different structures (unordered structure, α-helix, and β-sheet). Then, the authors assessed secondary structures of rituximab under thermal stress (up to 72°C) at pH 1.9. When the temperature is increased to 60°C and above, clear broadening of the amide I band was observed with a shift from 1670 to 1660 cm^{-1}, which indicates the unfolding of rituximab from its β-sheet structure to partially unordered structure. In another report, Gómez et al. applied Raman spectroscopy and perturbation-correlation moving windows to study mAb aggregation (Gómez de la Cuesta, Goodacre, & Ashton, 2014). In this study, Gómez et al. acquired Raman spectra of five mAb samples (one control and four mutants with different aggregate propensities). These samples were thermally stressed under 56°C–78°C. The excitation wavelength was 532 nm. This study showed that Raman spectroscopy is a sensitive technique to monitor mAb structural changes and provide insightful information about aggregation mechanisms, that is, it can differentiate between protein unfolding and aggregation. The change in peak intensity of Trp residues during the heating process revealed Fab domain unfolding, Fc domain unfolding, and mAb aggregation.

In summary, Raman spectroscopy is a noninvasive method and has a broad application to biotherapeutic development. The high sensitivity and reproducibility and insensitive to water make this method suitable for protein structural characterization, identity test, and real-time monitoring of protein manufacturing (Abu-Absi et al., 2011). While both Raman and FTIR spectroscopies are vibrational spectroscopy, these two techniques differ in some key features: Raman scattering is a two-photon event and measures the shift of the radiation wavelength, whereas IR absorption is a one-photon event and measures the absolute absorbance of the radiation. In addition, Raman scattering is sensitive to the polarizability change of a molecule, whereas IR absorption is sensitive to the dipole moment change of a molecule. As a result, homonuclear chemical bonds such as C—C, C = C, and C≡C bonds are Raman sensitive, whereas heteronuclear chemical bonds, for example, O—H bonds in water molecules, are IR sensitive. Therefore Raman spectroscopy is complimentary to the FTIR technique.

6.4 Fluorescence spectroscopy

Fluorescence spectroscopy is based on the fluorescent property of a molecule. When electrons of a molecule absorb the energy from the photon of the light in its absorption wavelength range, the electrons jump from the ground state to an excited state for a short period of time, then return to the ground state and release the energy by emitting a new photon in a longer wavelength and heat (Lakowicz, 2013; Sauer, Hofkens, & Enderlein, 2011). Therefore fluorescence phenomenon is a two one-photon events, that is, photon absorption followed by photon emission. Typically the time that the molecule spends in the excited state, that is, fluorescence lifetime, is in the scale of 10^{-12} to 10^{-9} s (Itagaki, 2000). Though an electron typically reserves its spin direction when it is excited to a singlet state, that is, the chemical bond with an even number of electrons still has two electrons with opposite spin directions, the spin direction of the excited electron can be reversed through intersystem crossing. In this case, the electron is excited to a triplet state. When the electron returns from the triplet state to the ground state, the emission is called phosphorescence. The lifetime of phosphorescence is much longer than that of fluorescence, which is typically in the scale of 10^{-4} to 10^{2} s. In this chapter, we will focus on the application of fluorescence on mAb HOS characterization.

Because fluorescence is very sensitive to its local environment, for example, polarity, nearby quencher, it has been widely used for protein structure characterization (Basak & Chattopadhyay, 2014). During the fluorescence measurements, emission peak intensity and maximum wavelength, emission peak width, and the fluorescence lifetime can provide valuable information to reveal protein folding (tertiary structure), protein–protein or protein–ligand interactions, protein thermal stability, etc. Huang, Hsieh, and Zhou (2017). Depending on the type of fluorophores, fluorescence can be classified as intrinsic and extrinsic. Intrinsic fluorescence can be observed when molecules contain natural fluorophores, such as aromatic amino acids in proteins. Extrinsic fluorescence can be measured when an external fluorescent dye is added to the molecule, such as 8-anilino-1-naphthalenesulfonic acid (ANS), Nile Red, or SYPRO Orange dye (Demeule, Palais, Machaidze, Gurny, & Arvinte, 2009; Gabellieri & Strambini, 2006; Hawe, Sutter, & Jiskoot, 2008; Zheng, Bantog, & Bayer, 2011). These extrinsic fluorescence dyes do not have strong fluorescence emission in the aqueous solution. However, when they are in a hydrophobic microenvironment or bind to a hydrophobic region of proteins, their fluorescence intensity can increase substantially, and the emission peak will shift to a lower wavelength, that is, a blue shift.

For intrinsic fluorescence, although all aromatic residues are natural fluorophores in proteins, Trp is predominantly used due to its strong absorbance of the UV light around 280 nm by its indole ring (Ghisaidoobe & Chung, 2014; Teale & Weber, 1957). Researchers often use 295 nm as the excitation wavelength for protein measurements to further reduce the fluorescence contributions from Phe and Tyr residues. As a hydrophobic residue, Trp is typically buried inside a protein's hydrophobic core. The emission peak position and intensity of Trp are highly dependent on its local environment (Vivian & Callis, 2001). When the local environment is more hydrophobic, the Trp emission peak can exhibit a blue shift, that is, maximum emission (λ_{EM}) moves to a shorter wavelength region. If the local environment becomes more polar, the Trp emission peak moves to a longer wavelength region as a red shift with a reduced intensity due to reduced quantum yield (Zheng, Middaugh, & Siahaan, 2009). It has been proposed that the λ_{EM} of Trp in a protein can vary from ~308 nm, where Trp is buried in a hydrophobic core (Finazzi-Agro et al., 1970), to ~350 nm, where Trp is fully exposed to solvent (Eftink, 2002). In Liu's work, shifts of the Trp emission peak was used to assess impact of freezing stress on the tertiary structure of an IgG2 mAb (Liu, Braun, Wang, Randolph, & Carpenter, 2014). The freezing process can induce protein aggregation. The authors attempted to use the two-step theory (Chi, Krishnan, Randolph, & Carpenter, 2003), protein tertiary structural change followed by reduced colloidal stability, to explore the aggregation mechanism of the mAb in various formulations. Although the emission peak cannot differentiate the microenvironments of individual Trp residues as the studied mAb contains 24 Trp residues, it can still be used to monitor changes of the overall tertiary structure of the mAb. The authors observed the changes of the mAb tertiary structure in the frozen state based on the shifts of the average emission peak and concluded that the level of emission peak shifts did not correlate with the level of mAb aggregation. Although the emission spectra itself cannot provide high-resolution information of the protein HOS as proteins often contain multiple Trp residues, this report showed that intrinsic fluorescence can be used as a simple and effective method to investigate the aggregation mechanism of mAbs under the freeze/thaw stress.

Because Trp emission of proteins is typically a broad peak and represents an averaged emission of several Trp residues with poor resolution, it cannot provide characteristic information other than peak intensity, λ_{EM}, and peak width. Therefore various mathematical methods have been applied to achieve a better resolution and enhanced structure information (Abbas, Gaspar, Sharma, Patapoff, & Kalonia, 2013; Huang et al., 2017). One approach is to calculate the second derivative of a fluorescence spectrum to improve its resolution and resolve shoulder peaks, which is similar to the data processing method used in FTIR. Abbas et al. applied this method to characterize subtle structural changes of a mAb under various stress conditions (Abbas et al., 2013). After calculating second derivatives of Trp emission spectra, the authors identified three characteristic bands: class I at around 330 nm, class II at around 340 nm, and class III at around 350 nm. From class I to class III, it represents a trend that the microenvironment of Trp residues changes from a relatively polar but a rigid environment to more surface exposure and in more contact with free water. Thus protein unfolding can be shown as increased intensity in the class III band with decreased intensity in the class I and II bands. In another report (Huang et al., 2017), Huang et al. used a mathematical log-normal modeling method to enhance the resolution of fluorescence spectra and achieved the quantitative measurement. Fluorescence spectra of four IgG1 mAbs were collected. After using the log-normal modeling to calculate the emission maximum and emission bandwidth, the authors observed clear similarities and differences for the tested mAbs in a 2D plot of emission bandwidth versus emission maximum. The authors further applied this method to manufacturing comparability assessment, where mAb2 from two different manufacturing processes were evaluated. A total of six batches (three batches for each process) were tested. The bandwidth-maximum plot supports that the materials from the two different processes are comparable from the HOS perspective. At the end of the report, the authors also explored the use of this method to monitor the structural changes of mAb2 under guanidine hydrochloride (GdnHCl) denaturation. Overall, Huang et al. showed that intrinsic Trp fluorescence can be used as a quantitative method to reveal changes of mAb HOS during mAb formulation development and manufacturing.

Besides intrinsic fluorescence, researchers also used extrinsic fluorescence to characterize protein HOS as the extrinsic fluorescence dye is sensitive to the polarity of the microenvironment. Gabellieri et al. used the ANS fluorescent probe to understand perturbation of protein tertiary structure due to freezing. They studied several proteins in different sizes and complexity in order to draw a general conclusion (Gabellieri & Strambini, 2003; Gabellieri & Strambini, 2006). Their findings showed that ANS binding is sensitive to protein tertiary structural changes. In addition, temperature of ice and conditions affecting the globular structure stability can impact ANS binding and the corresponding fluorescence signal. Their data showed that the protein tertiary structure was perturbed when reducing the temperature from $-4°C$ to $-18°C$. The level of ANS binding correlates with the magnitude of protein structure perturbation, that is, the more exposure of protein's hydrophobic regions, the stronger binding of ANS to proteins. In another report, Demeule et al. used ANS along with other analytical methods to assess trastuzumab aggregation after diluting in either 0.9% NaCl or 5% dextrose infusion solutions (Demeule et al., 2009). As mentioned previously, protein aggregation is due to partial unfolding of the native structure, that is, formation of molten globules (Chi et al., 2003). ANS is a sensitive probe to monitor the partial unfolding of proteins. In this work, besides collecting ANS emission spectra, Demeule et al. also measured fluorescence anisotropy and lifetime to confirm ANS binding to trastuzumab. The data showed stronger ANS fluorescence and a blue shift in the emission spectrum when trastuzumab was diluted in the 5% dextrose solution than in the 0.9% NaCl solution. This result indicated a higher level of trastuzumab aggregation in the 5% dextrose solution. The higher anisotropy and longer fluorescence lifetime of ANS in the 5% dextrose solution with diluted trastuzumab further confirmed this result.

In summary, fluorescence spectroscopy is a sensitive and powerful method for mAb HOS characterization with various advantages: it requires a very small amount of sample, offers a high S/N ratio, allows rapid data acquisition, and can be used with both intrinsic fluorophore (e.g., Trp) and extrinsic fluorophore (e.g., ANS, Nile Red, or SYPRO Orange).

6.5 Free thiol analysis

Free thiol is a common posttranslational modification of proteins (Maulik, Jennifer, & Teruna, 2009). Varying levels of free thiols at different locations have been observed in multiple therapeutic mAbs. Various analytical methods have been developed to detect and quantify free thiols in a protein providing either total amount or site-specific information.

Total free thiol content at the intact molecular level can be measured with spectroscopy-based methods (colorimetric or fluorescent) by reacting with chromogenic or fluorescent probes. Ellman's reagent, DTNB, was first introduced in 1959 for the estimation of free thiol groups (Ellman, 1959) and is the most widely used method for the quantitation of thiols in peptides and proteins. DTNB is an aromatic disulfide that reacts with thiols by an exchange reaction to form a protein conjugate that contains one molecule of the chromogenic product, 5-thio-2-nitrobezoic acid (TNB). Due to the low pK_a of the TNB at 4.5, it forms an extremely efficient leaving group. Upon release, TNB then ionizes to the TNB^{2-} dianion in water, producing a yellow color. Detection of free thiols is achieved by monitoring the absorbance of the TNB^{2-} at 412 nm using a spectrophotometer. Levels of free thiols in protein samples may be estimated by comparison to a standard curve composed of known concentrations of a thiol-containing compound such as cysteine or glutathione (GSH). Alternatively, free thiols may be quantitated by reference to the extinction coefficient of TNB^{2-}. While DTNB has weak absorption at 412 nm, the extinction coefficient for TNB^{2-} is 14,100 $M^{-1}cm^{-1}$ at pH 7.3 (Riddles, Blakeley, & Zerner, 1983). This value, however, drops sharply at pH below protonation of the orange thiolate species. In addition, it can vary depending on ionic strength. For example, the extinction coefficient of TNB^{2-} at 412 nm decreases to 13,700 M^{-1} cm^{-1} when thiols are determined in 6 M GdnHCl (Winther & Thorpe, 2014).

There have been many reports of using Ellman's assay for the quantitation of thiols in mAbs. For example, it was found that every mole of omalizumab (an anti-IgE antibody) contains 0.9 mole of free thiols, which resulted in a significant decrease in potency (Harris, 2005). Although fast and highly versatile, there are numerous disadvantages of DTNB. The rate of the reaction is highly dependent on the pH and the pK_a of the thiol group, which means a reaction pH of at least 7.0, is normally required. However, at this pH, DTNB may undergo hydrolysis at elevated temperatures, giving rise to varying background absorbance over time. Other unwanted side reactions such as alkylation and oxidation reactions may also occur to the thiolate anion, potentially leading to an underestimate of the free thiol level in the sample. The most significant drawback is its limited sensitivity with a detection limit of ~3 μM, making it challenging for free thiol detection when sample amount is limited.

There have been extensive efforts in exploring the use of fluorescent labels to improve the sensitivity of spectroscopic free thiol detection. Due to the strong nucleophilicity of the thiol group, it can react readily with electrophiles such as Michael acceptors and alkylating agents (Koval', 2007). Reaction of thiols with maleimides via Michael addition is particularly popular for bioconjugation and labeling of proteins for thiol detection. A series of maleimide-based fluorescent labels were applied in studying antibody free thiols (Rusinova, Tretyachenko-Ladokhina, Vele, Senear, & Ross, 2002). Zhang and Czupryn (2002) reported the use of a fluorescent probe *N*-(1-pyrenyl) maleimide (NPM) in determining the free thiol content in purified recombinant IgG1, IgG2, and IgG4 antibodies. NPM was selected due to its specificity, sensitivity, and ease of use. In particular, NPM is essentially nonfluorescent until the free thiol group is conjugated, which allows measurement of the fluorescence of the sample without prior reagent separation. With the excitation wavelength set at 330 nm, the fluorescence emission intensities of the NPM-derivatized mAb samples were monitored at 380 nm and compared to a series of Ac-Cys standards. Free thiol levels in all the IgG samples were successfully quantified with this approach under both native and denaturing conditions.

Although fluorescence-based spectroscopic detection of free thiols has gained tremendous attentions due to its inherent advantages of high sensitivity and selectivity, many of these probes have drawbacks that limit their applications in detecting free thiols in proteins. A particular challenge associated with these hydrophobic fluorescent probes is limited water solubility. In addition, there is potential fluorescence interference from excess amount of labeling agent, which may be mitigated if the labeling agent show significantly increased signal after conjugation. In some cases, the labeled protein product is unstable, which can lead to inaccuracy in the quantitative determination of the thiol group. Lastly, fluorescent response of the labeled protein is often substrate-dependent and quantitation of free thiols often requires the use of standards identical to the substrates.

Even with the limitations of DTNB or fluorescent probes described above, it is still relatively convenient to measure total free thiols than obtaining site-specific free thiol information during the development of mAb therapeutics. In general, higher levels of total free thiols are observed under denaturing conditions than native conditions. Huh et al. (2013) showed the amount of free thiols accessible to DTNB was found to be negligible in the native mAbs. With the more sensitive detection using NPM, Zhang and Czupryn (2002) showed approximately 0.02 mol of free thiol per mole of protein under native conditions. Upon denaturation, minor species of the mAbs were observed on sodium dodecyl sulfate polyacrylamide gel electrophoresis (SDS-PAGE), and the free thiol level was determined to be approximately 0.1 mol/mol of protein. Similarly, Huh et al. reported that free thiols in the range of 0.17–0.59 mol/mol of protein were found in all recombinant antibodies studied under the denaturing condition with 5 M GdnHCl present. This general observation of higher free thiol level in the denatured mAbs indicates that free thiols also reside within buried regions of antibodies in the hydrophobic domains.

During early development phase, spectroscopic methods such as Ellman's reaction that provide a direct measurement of the total free thiol level are often applied in the characterization of therapeutic mAbs. Some of these methods may be developed in a high-throughput fashion to support candidate screening and even process characterization studies. However, in order to assess the impact of free thiols on protein structural and functional integrity, site-specific (or domain-specific) detection and quantification is required. This is particularly important for late stage development where comprehensive protein characterization needs to be performed. Separation techniques such as chromatographic or electrophoretic methods are required to differentiate various species containing free thiols at different locations. Chromatographic methods such as IEC (Ouellette et al., 2010) or hydrophobic interaction chromatography (HIC) (Harris, 2005) were used to monitor free thiol species with incomplete formation of the intrachain disulfide bond in the heavy chain variable domain (V_H) in IgG1 mAbs. Cheng et al. (2017) reported using *N-tert*-butylmaleimide (NtBM) alkylation followed by reversed-phase chromatography (RP-HPLC) and online MS for rapid total and site-specific free thiol characterization of IgG1. However, further characterization by peptide mapping with LC-MS analysis is often required to identify the specific location of free thiols in the collected fractions of variants. Differential isotope labeling of cysteine residues followed by peptide mapping with LC-MS analysis is a powerful tool to provide simultaneous detection and quantification of site-specific free thiols (Xiang et al., 2009). Free thiol group was first alkylated with ^{12}C iodoacetic acid (IAA), whereas the other thiols generated from the reduction of disulfide bonds was labeled with ^{13}C IAA. Cysteine-containing peptides that were modified by ^{13}C IAA showed a MW that was 2 Da higher than the same peptide that was modified by ^{12}C IAA. Peptides, containing the same cysteine residues that were modified with both alkylation reagents, coeluted on RP-HPLC. Analysis by MS resulted in two partially overlapped *m/z* series for each cysteine containing peptide, corresponding to modification by IAA with ^{12}C or ^{13}C. The percentage of free thiol was then calculated using the two *m/z* series at each cysteine site. With this approach, low levels of free thiols at every cysteine residue in four mAbs were detected.

There are 12 intrachain disulfide bonds and 4 interchain disulfide bonds in an IgG. Interestingly free thiols were detected at all of these locations in the antibodies. At the interchain disulfide linkage locations, free thiol levels were found to be the lowest, typically below 1% (Chumsae, Gaza-Bulseco, & Liu, 2009; Xiang et al., 2009). These free thiols are likely present at the end of the protein folding and assembly process or generated due to the reduction by thioredoxin or GSH reductase during cell culture manufacturing. In addition to the LC-MS peptide mapping method, free thiol species due to reduction of the interchain disulfide bonds will appear as SDS-dissociable fragments such as LCs, HCs, or half antibodies in the nonreduced capillary electrophoresis sodium dodecyl sulfate (CE-SDS) analysis. In general, low levels of free thiols at the interchain disulfide bonds generated from reduction under mild and nondenaturing conditions are thought to have a limited effect on antibody structure and stability. That is because assembly of light and heavy chains does not depend primarily on covalent disulfide linkages, but rather on noncovalent interactions. Significant reduction of the interchain disulfide bonds, however, will disrupt the mAb structure, resulting in elevated fragment levels, loss of potency and/or increased levels of aggregates during manufacturing and storage (Van Buren, Rehder, Gadgil, Matsumura, & Jacob, 2009).

Different from interchain disulfide bonds, intrachain disulfide bonds are generally buried between the two layers of antiparallel β-sheet structures within each domain and are not solvent-exposed. Therefore free thiols associated with intrachain disulfide bonds are often detected only under denaturing conditions. In the constant domains (C_L, C_H1, C_H2, and C_H3) of recombinant IgGs, free thiols are generally found at low levels (Brych et al., 2010; Lacy et al., 2008; Liu, Gaza-Bulseco, Chumsae, & Newby-Kew, 2007; Zhang, Marzilli, Rouse, & Czupryn, 2002). Up to 7.5% free thiols in the constant domains were observed in commercially available therapeutic IgG1 antibodies (Huh et al., 2013). In addition, it was reported that human sera-derived IgG1 and IgG2 antibodies contain free thiols in the range of 1%−4% in the constant domains (Huh et al., 2013). It is generally believed that these intrachain free thiols in the constant domains do not pose a significant impact on the overall mAb HOS. Several studies showed intrachain disulfide bonds are not necessary for proper folding of Ig domains (Frisch et al., 1996; Robinson, Pringle, Woolhead, & Bulleid, 2017) and functional Ig domains can exist without disulfides (Rudikoff & Pumphrey, 1986). Additional studies showed individual domains of C_L (Kikuchi, Goto, & Hamaguchi, 1986) and C_H3 (Mcauley et al., 2008) without the complete intrachain disulfide bonds had no substantial structural changes compared to the disulfide-bonded forms.

Free thiols at the intrachain disulfide linkage location in the variable domains (V_H and V_L), however, are shown to be much more variable across products, batches, or processes. For example, higher percentages of free thiols localized in the V_H domain were detected in omalizumab (at 31%, internal data) and ocrelizumab (at 15%−19%) (Zhang et al., 2012), whereas majority of other mAbs showed free thiol below 5% (internal data). The impact of free thiols in the V_H or V_L domain on the overall antibody HOS or even antigen binding needs to be carefully assessed because of their close proximity to the complementarity-determining regions. In the case of omalizumab, elevated levels of free thiols were initially discovered using Ellman's reagent under native conditions (Harris, 2005). Detailed characterization of the molecule revealed free thiols at positions Cys22 and Cys96 in the V_H domain at up to 31%. Free thiol-containing Fab was then enriched by papain digestion followed by chromatographic purification with HIC. A substantial reduction (~70%) in potency was observed for the free thiol-containing Fab in the inhibition of IgE-receptor binding assay. In 2012 Zhang et al. (2012) reported elevated levels of free thiols in ocrelizumab also at positions Cys22 and Cys96 in the V_H domain of the IgG1 molecule. However, in this case, the free thiol-containing Fab and the corresponding disulfide-bonded form exhibited comparable antigen-binding activities to the target, CD20. In addition, full-length ocrelizumab containing the V_H free thiols (predominantly in the heterodimer form containing free thiol in only one chain) was enriched by cation-exchange chromatography. It showed equal activities (106% CDC and 92% ADCC normalized to the activities of the control, respectively). Unlike omalizumab, these free thiols in the V_H

domain of ocrelizumab were found to be buried in the hydrophobic domain and could not be detected by Ellman's reaction under native conditions. Interestingly, there is evidence from rats that rapid conversion of V_H free thiols to disulfide bonds occurs in systemic circulation with intravenous administration (Ouellette et al., 2010). If free thiols are indeed rapidly converted to disulfide bonds within 24 h of administering a therapeutic antibody, the risk of a V_H free thiol-containing variant having a negative product quality impact may be low.

Free thiols have been considered as indicators for decreased antibody thermal stability (Lacy et al., 2008). Kao et al. reported (unpublished results) a decrease of 4°C in the melting temperature of the V_H free thiol-containing Fab compared to the disulfide-bonded form. Free thiols have been associated with covalent aggregation under some harsh stress conditions including agitation stress and high temperature and/or high pH treatment over a long period of time. Free thiols in the C_H1 domain of an IgG2 molecule have been associated with covalent aggregation under severe agitation stress (agitated at 500 rpm for 72 h at room temperature on an orbital shaker) (Huh et al., 2013). In an IgG1 Fab' material containing free thiols, the percent of dimers increased when the pH changed from 5.8 to 9.5 and the dimerization was shown to be mediated through the free thiol groups only (Kamat, Tolman, & Brown, 2002). In a separate study, it was shown that the majority of the dimer of an IgG1 molecule formed after incubation at 37°C over 5 months in PBS at pH 7.4 contained intermolecular disulfide bonds, which could result from free thiols (Remmele et al., 2006). A major aggregation pathway to form protein dimers for a recombinant monoclonal IgG2 antibody at pH 6.0 after heat stress (at 37°C for 15 weeks) is through disulfide bond formation (Van Buren et al., 2009).

However, impact of free thiols on mAb real-time stability is likely to be minimal under typical storage conditions (e.g., with typical formulation pH below 7) instead of harsh stress conditions. There is no evidence of stability issues related to the V_H free thiols in ocrelizumab (Zhang et al., 2012). High MW forms remain consistent through the end of shelf life of ocrelizumab Drug Product and free thiol levels did not change in the ocrelizumab Drug Product samples stored at 5°C for 3 years. Similarly, free thiol levels remained the same in lyophilized omalizumab Drug Product samples (containing up to 31% free thiol in the V_H domain) stored at 40°C for up to 12 weeks, or up to 8 weeks at 50 or 60°C (unpublished results). The unreactive nature of free thiols observed is consistent with findings from a study that showed purified human C_H3

domain in the reduced state was stable at pH 5.0 for several months at 4°C and did not undergo spontaneous disulfide bond formation by dissolved atmospheric oxygen (Mcauley et al., 2008).

Free thiol analysis may be complicated because it was found at every cysteine residue in the mAbs. To assess the impact of free thiols on protein structure, function, and stability, one needs to consider the location and levels of free thiols, mechanism of action and the route of administration of the therapeutic product, as well as the specific formulation and storage conditions. These considerations will guide the selection of appropriate analytical tools for the detection of free thiols in mAbs.

References

Abbas, S. A., Gaspar, G., Sharma, V. K., Patapoff, T. W., & Kalonia, D. S. (2013). Application of second-derivative fluorescence spectroscopy to monitor subtle changes in a monoclonal antibody structure. *Journal of Pharmaceutical Sciences, 102*(1), 52–61.

Abu-Absi, N. R., Kenty, B. M., Cuellar, M. E., Borys, M. C., Sakhamuri, S., Strachan, D. J., Hausladen, M. C., & Li, Z. J. (2011). Real time monitoring of multiple parameters in mammalian cell culture bioreactors using an in-line Raman spectroscopy probe. *Biotechnology and Bioengineering, 108*(5), 1215–1221.

Arbogast, L. W., Brinson, R. G., & Marino, J. P. (2015). Mapping monoclonal antibody structure by 2D 13C NMR at natural abundance. *Analytical Chemistry, 87*(7), 3556–3561.

Arbogast, L. W., Delaglio, F., Schiel, J. E., & Marino, J. P. (2017). Multivariate analysis of two-dimensional 1H, 13C methyl NMR spectra of monoclonal antibody therapeutics to facilitate assessment of higher order structure. *Analytical Chemistry, 89*(21), 11839–11845.

Asher, S. A., Bormett, R. W., Chen, X. G., Lemmon, D. H., Cho, N., Peterson, P., Arrigoni, M., Spinelli, L., & Cannon, J. (1993). UV resonance Raman spectroscopy using a new cw laser source: convenience and experimental simplicity. *Applied Spectroscopy, 47*(5), 628–633.

Asher, S. A., Ludwig, M., & Johnson, C. R. (1986). UV resonance Raman excitation profiles of the aromatic amino acids. *Journal of the American Chemical Society, 108*(12), 3186–3197.

Aubin, Y., Hodgson, D. J., Thach, W. B., Gingras, G., & Sauvé, S. (2015). Monitoring effects of excipients, formulation parameters and mutations on the high order structure of filgrastim by NMR. *Pharmaceutical Research, 32*(10), 3365–3375.

Baldassarre, M., Li, C., Eremina, N., Goormaghtigh, E., & Barth, A. (2015). Simultaneous fitting of absorption spectra and their second derivatives for an improved analysis of protein infrared spectra. *Molecules, 20*(7), 12599–12622.

Basak, S., & Chattopadhyay, K. (2014). Studies of protein folding and dynamics using single molecule fluorescence spectroscopy. *Physical Chemistry Chemical Physics, 16*(23), 11139–11149.

Bath, A., & Zscherp, C. (2002). What vibrations tell about proteins. *Quarterly Reviews of Biophysics, 35*, 369–430.

Bax, A. (1989). Two-dimensional NMR and protein structure. *Annual Review of Biochemistry, 58*(1), 223–256.

Berman, H. M., Bourne, P. E., Westbrook, J., & Zardecki, C. (2003). *The protein data bank. Protein Structure* (pp. 394–410). CRC Press.

Bloch, F. (1946). Physics today, Dec. 1976, 29, 23.[3] EM Purcell, HC Torrey, and RV Pound. *Physical Review, 69*, 37.

Brinson, R. G., Marino, J. P., Delaglio, F., Arbogast, L. W., Evans, R. M., Kearsley, A., Gingras, G., Ghasriani, H., Aubin, Y., Pierens, G. K., & Jia, X. (2019). Enabling adoption of 2D-NMR for the higher order structure assessment of monoclonal antibody therapeutics. In *MAbs* (Vol. 11, pp. 94–105). Taylor & Francis, No. 1.

Brych, S. R., Gokarn, Y. R., Hultgen, H., Stevenson, R. J., Rajan, R., & Matsumura, M. (2010). Characterization of antibody aggregation: Role of buried, unpaired cysteines in particle formation. *Journal of Pharmaceutical Sciences*, *99*(2), 764–781.

Bueno, J., Long, D., Kauffman, J. F., & Arzhantsev, S. (2015). Deep-ultraviolet resonance Raman (DUVRR) spectroscopy of therapeutic monoclonal antibodies subjected to thermal stress. *Analytical Chemistry*, *87*(15), 7880–7886.

Byler, D. M., & Susi, H. (1986). Examination of the secondary structure of proteins by deconvolved FTIR spectra. *Biopolymers: Original Research on Biomolecules*, *25*(3), 469–487.

Carey, P.R., (1982). Molecular biology: biochemical applications of Raman and resonance Raman spectroscopies.

Chalmers, J., & Griffiths, P. (2002). *Handbook of Vibrational Spectroscopy, 5 volumes set*. Wiley.

Chen, K., Freedberg, D. I., & Keire, D. A. (2015). NMR profiling of biomolecules at natural abundance using 2D 1H–15N and 1H–13C multiplicity-separated (MketoS) HSQC spectra. *Journal of Magnetic Resonance*, *251*, 65–70.

Chen, K., Long, D. S., Lute, S. C., Levy, M. J., Brorson, K. A., & Keire, D. A. (2016). Simple NMR methods for evaluating higher order structures of monoclonal antibody therapeutics with quinary structure. *Journal of Pharmaceutical and Biomedical Analysis*, *128*, 398–407.

Chen, Y. J., Hsieh, Y. W., Cheng, Y. D., & Liao, C. C. (2001). Study on the secondary structure of protein in amide I band from human colon cancer tissue by Fourier-transform infrared spectroscopy. *Chang Gung Medical Journal*, *24*(9), 541–546.

Cheng, Y., Chen, M. T., Patterson, L. C., Yu, X. C., Zhang, Y. T., Burgess, B. L., & Chen, Y. (2017). Domain-specific free thiol variant characterization of an IgG1 by reversed-phase high-performance liquid chromatography mass spectrometry. *Analytical Biochemistry*, *519*, 8–14.

Chi, E. Y., Krishnan, S., Randolph, T. W., & Carpenter, J. F. (2003). Physical stability of proteins in aqueous solution: mechanism and driving forces in nonnative protein aggregation. *Pharmaceutical Research*, *20*(9), 1325–1336.

Cho, N., & Asher, S. A. (1993). UV resonance Raman studies of DNA-pyrene interactions: optical decoupling Raman spectroscopy selectively examines external site bound pyrene. *Journal of the American Chemical Society*, *115*(14), 6349–6356.

Chumsae, C., Gaza-Bulseco, G., & Liu, H. (2009). Identification and localization of unpaired cysteine residues in monoclonal antibodies by fluorescence labeling and mass spectrometry. *Analytical Chemistry*, *81*(15), 6449–6457.

Cooley, J. W., & Tukey, J. W. (1965). An algorithm for the machine calculation of complex Fourier series. *Mathematics of Computation*, *19*(90), 297–301.

Correia, I. (2010). Stability of IgG isotypes in serum. In *MAbs* (Vol. 2, pp. 221–232). Taylor & Francis, No. 3.

Cressey, D., & Callaway, E. (2017). Cryo-electron microscopy wins chemistry Nobel. *Nature News*, *550*(7675), 167.

Demeule, B., Palais, C., Machaidze, G., Gurny, R., & Arvinte, T. (2009). New methods allowing the detection of protein aggregates: A case study on trastuzumab. In *MAbs* (Vol. 1, pp. 142–150). Taylor & Francis, No. 2.

Deshpande, S. S. (2002). *Handbook of food toxicology*. CRC Press.

Dong, A., Huang, P., & Caughey, W. S. (1990). Protein secondary structures in water from second-derivative amide I infrared spectra. *Biochemistry*, *29*(13), 3303–3308.

Dong, A., Huang, P., & Caughey, W. S. (1992). Redox-dependent changes in beta-extended chain and turn structures of cytochrome c in water solution determined by second derivative amide I infrared spectra. *Biochemistry*, *31*(1), 182–189.

Dousseau, F., & Pezolet, M. (1990). Determination of the secondary structure content of proteins in aqueous solutions from their amide I and amide II infrared bands. Comparison between classical and partial least-squares methods. *Biochemistry, 29*(37), 8771–8779.

Dreker, L., Schwarz, J., Reichel, W., & Hilschmann, N. (1976). Rule of antibody structure. The primary structure of a monoclonal IgG1 immunoglobulin (myeloma protein Nie), I: Purification and characterization of the protein, the L-and H-chains, the cyanogenbromide cleavage products, and the disulfide bridges (author's transl). *Hoppe-Seyler's Zeitschrift fur physiologische Chemie, 357*(11), 1515–1540.

Eftink, M. R. (2002). *Intrinsic fluorescence of proteins. Topics in fluorescence spectroscopy* (pp. 1–15). Boston, MA: Springer.

Ellman, G. L. (1959). Tissue sulfhydryl groups. *Archives of Biochemistry and Biophysics, 82*(1), 70–77.

Englander, S. W. (2006). Hydrogen exchange and mass spectrometry: A historical perspective. *Journal of the American Society for Mass Spectrometry, 17*(11), 1481–1489.

Ettah, I., & Ashton, L. (2018). Engaging with Raman spectroscopy to investigate antibody aggregation. *Antibodies*, 7(3), 24.

Farrar, T. C., & Becker, E. D. (2012). *Pulse and Fourier transform NMR: Introduction to theory and methods*. Elsevier.

FDA. (2015). *Guidance for industry: Scientific considerations in demonstrating biosimilarity to a reference product. Maryland: FDA.*

Feige, M. J., Hendershot, L. M., & Buchner, J. (2010). How antibodies fold. *Trends in Biochemical Sciences, 35*(4), 189–198.

Ferguson, C. N., & Gucinski-Ruth, A. C. (2016). Evaluation of ion mobility-mass spectrometry for comparative analysis of monoclonal antibodies. *Journal of the American Society for Mass Spectrometry, 27*(5), 822–833.

Fernández, C., & Wider, G. (2003). TROSY in NMR studies of the structure and function of large biological macromolecules. *Current Opinion in Structural Biology, 13*(5), 570–580.

Finazzi-Agro, A., Rotilio, G., Avigliano, L., Guerrieri, P., Boffi, V., & Mondovi, B. (1970). Environment of copper in Pseudomonas fluorescens azurin: fluorometric approach. *Biochemistry, 9*(9), 2009–2014.

Foster, M. P., McElroy, C. A., & Amero, C. D. (2007). Solution NMR of large molecules and assemblies. *Biochemistry, 46*(2), 331–340.

Franks, J., Glushka, J. N., Jones, M. T., Live, D. H., Zou, Q., & Prestegard, J. H. (2016). Spin diffusion editing for structural fingerprints of therapeutic antibodies. *Analytical Chemistry, 88*(2), 1320–1327.

Frisch, C., Kolmar, H., Schmidt, A., Kleemann, G., Reinhardt, A., Pohl, E., Usón, I., Schneider, T. R., & Fritz, H. J. (1996). Contribution of the intramolecular disulfide bridge to the folding stability of REIv, the variable domain of a human immunoglobulin κ light chain. *Folding and Design, 1*(6), 431–440.

Gabellieri, E., & Strambini, G. B. (2003). Perturbation of protein tertiary structure in frozen solutions revealed by 1-anilino-8-naphthalene sulfonate fluorescence. *Biophysical Journal, 85*(5), 3214–3220.

Gabellieri, E., & Strambini, G. B. (2006). ANS fluorescence detects widespread perturbations of protein tertiary structure in ice. *Biophysical Journal, 90*(9), 3239–3245.

Gabrielson, J. P., Brader, M. L., Pekar, A. H., Mathis, K. B., Winter, G., Carpenter, J. F., & Randolph, T. W. (2007). Quantitation of aggregate levels in a recombinant humanized monoclonal antibody formulation by size-exclusion chromatography, asymmetrical flow field flow fractionation, and sedimentation velocity. *Journal of Pharmaceutical Sciences, 96*(2), 268–279.

Gao, X., Bain, K., Bonanno, J. B., Buchanan, M., Henderson, D., Lorimer, D., Marsh, C., Reynes, J. A., Sauder, J. M., Schwinn, K., & Thai, C. (2005). High-throughput limited proteolysis/mass spectrometry for protein domain elucidation. *Journal of Structural and Functional Genomics*, *6*(2-3), 129–134.

Garcia, N. K., Deperalta, G., & Wecksler, A. T. (2019). Current trends in biotherapeutic higher order structure characterization by irreversible covalent footprinting mass spectrometry. *Protein and Peptide Letters*, *26*(1), 35–43.

Gevondyan, N. M., Volynskaia, A. M., & Gevondyan, V. S. (2006). Four free cysteine residues found in human IgG1 of healthy donors. *Biochemistry (Moscow)*, *71*(3), 279–284.

Ghisaidoobe, A. B., & Chung, S. J. (2014). Intrinsic tryptophan fluorescence in the detection and analysis of proteins: A focus on Förster resonance energy transfer techniques. *International Journal of Molecular Sciences*, *15*(12), 22518–22538.

Gómez de la Cuesta, R., Goodacre, R., & Ashton, L. (2014). Monitoring antibody aggregation in early drug development using Raman spectroscopy and perturbation-correlation moving windows. *Analytical Chemistry*, *86*(22), 11133–11140.

Goormaghtigh, E., Ruysschaert, J. M., & Raussens, V. (2006). Evaluation of the information content in infrared spectra for protein secondary structure determination. *Biophysical Journal*, *90*(8), 2946–2957.

Greenfield, N. J. (2006). Using circular dichroism spectra to estimate protein secondary structure. *Nature protocols*, *1*(6), 2876.

Griffiths, P. R., & De Haseth, J. A. (2007). *Fourier transform infrared spectrometry* (Vol. 171). John Wiley & Sons.

Haberger, M., Leiss, M., Heidenreich, A. K., Pester, O., Hafenmair, G., Hook, M., Bonnington, L., Wegele, H., Haindl, M., Reusch, D., & Bulau, P. (2016). Rapid characterization of biotherapeutic proteins by size-exclusion chromatography coupled to native mass spectrometry. In *MAbs* (Vol. 8, pp. 331–339). Taylor & Francis, No. 2.

Haris, P. I., & Severcan, F. (1999). FTIR spectroscopic characterization of protein structure in aqueous and non-aqueous media. *Journal of Molecular Catalysis B: Enzymatic*, *7* (1-4), 207–221.

Harris, L. J., Larson, S. B., Hasel, K. W., & McPherson, A. (1997). Refined structure of an intact IgG2a monoclonal antibody. *Biochemistry*, *36*(7), 1581–1597.

Harris, L. J., Skaletsky, E., & McPherson, A. (1998). Crystallographic structure of an intact IgG1 monoclonal antibody. *Journal of Molecular Biology*, *275*(5), 861–872.

Harris, R.J., (2005). Heterogeneity of Recombinant Antibodies: linking Structure to Function.

Hawe, A., Sutter, M., & Jiskoot, W. (2008). Extrinsic fluorescent dyes as tools for protein characterization. *Pharmaceutical Research*, *25*(7), 1487–1499.

Holzgrabe, U. (2010). Quantitative NMR spectroscopy in pharmaceutical applications. *Progress in Nuclear Magnetic Resonance Spectroscopy*, *57*(2), 229–240.

Huang, C. Y., Hsieh, M. C., & Zhou, Q. (2017). Application of tryptophan fluorescence bandwidth-maximum plot in analysis of monoclonal antibody structure. *AAPS PharmSciTech*, *18*(3), 838–845.

Huh, J. H., White, A. J., Brych, S. R., Franey, H., & Matsumura, M. (2013). The identification of free cysteine residues within antibodies and a potential role for free cysteine residues in covalent aggregation because of agitation stress. *Journal of Pharmaceutical Sciences*, *102*(6), 1701–1711.

Itagaki, H. (2000). *Fluorescence spectroscopy. Experimental methods in polymer science* (pp. 155–260). *San Diego, CA: Academic Press.*

Japelj, B., Ilc, G., Marušič, J., Senčar, J., Kuzman, D., & Plavec, J. (2016). Biosimilar structural comparability assessment by NMR: From small proteins to monoclonal antibodies. *Scientific Reports*, *6*(1), 1–12.

Jiang, Y., Li, C., Nguyen, X., Muzammil, S., Towers, E., Gabrielson, J., & Narhi, L. (2011). Qualification of FTIR spectroscopic method for protein secondary structural analysis. *Journal of Pharmaceutical Sciences*, *100*(11), 4631–4641.

Johnson, C. M. (2013). Differential scanning calorimetry as a tool for protein folding and stability. *Archives of Biochemistry and Biophysics*, *531*(1-2), 100–109.
Kal'nin, N. N., & Ven'yaminov, S. Y. (1988). Quantitative measurement of the IR spectra of water solutions. *Journal of Applied Spectroscopy*, *49*(4), 1028–1032.
Kamat, M. S., Tolman, G. L., & Brown, J. M. (2002). *Formulation development of an antifibrin monoclonal antibody radiopharmaceutical. Formulation, characterization, and stability of protein drugs: Case histories* (pp. 343–364). Boston, MA: Springer.
Kato, K., Yamaguchi, Y., & Arata, Y. (2010). Stable-isotope-assisted NMR approaches to glycoproteins using immunoglobulin G as a model system. *Progress in Nuclear Magnetic Resonance Spectroscopy*, *4*(56), 346–359.
Khan, T., Kandola, T. S., Wu, J., Ketter, E., Venkatesan, S., Lange, J. J., Gama, A. R., Box, A., Unruh, J. R., Cook, M., & Halfmann, R. (2017). Quinary structure kinetically controls protein function and dysfunction. *BioRxiv*, 205690.
Khattari, Z. (2017). A correlation between secondary structure and rheological properties of low-density lipoproteins at air/water interfaces. *Journal of Biological Physics*, *43*(3), 381–395.
Kheddo, P., Cliff, M. J., Uddin, S., van der Walle, C. F., & Golovanov, A. P. (2016). Characterizing monoclonal antibody formulations in arginine glutamate solutions using 1H NMR spectroscopy. In *Mabs* (Vol. 8, pp. 1245–1258). Taylor & Francis, No. 7.
Kikuchi, H., Goto, Y., & Hamaguchi, K. (1986). Reduction of the buried intrachain disulfide bond of the constant fragment of the immunoglobulin light chain: Global unfolding under physiological conditions. *Biochemistry*, *25*(8), 2009–2013.
Kiss, R., Fizil, Á., & Szántay, C., Jr (2018). What NMR can do in the biopharmaceutical industry. *Journal of Pharmaceutical and Biomedical Analysis*, *147*, 367–377.
Kong, J., & Yu, S. (2007). Fourier transform infrared spectroscopic analysis of protein secondary structures. *Acta Biochimica et Bophysica Sinica*, *39*(8), 549–559.
Koval', I. V. (2007). Reactions of thiols. *Russian Journal of Organic Chemistry*, *43*, 319–346.
Krimm, S., & Bandekar, J. (1986). Vibrational spectroscopy and conformation of peptides, polypeptides, and proteins. In *Advances in protein chemistry* (Vol. 38, pp. 181–364). Academic Press.
Krishnan, R. S., & Shankar, R. K. (1981). Raman effect: History of the discovery. *Journal of Raman Spectroscopy*, *10*(1), 1–8.
Lacy, E. R., Baker, M., & Brigham-Burke, M. (2008). Free sulfhydryl measurement as an indicator of antibody stability. *Analytical Biochemistry*, *382*(1), 66–68.
Lakowicz, J. R. (Ed.), (2013). *Principles of fluorescence spectroscopy*. Springer Science & Business Media.
Lee, D. C., Haris, P. I., Chapman, D., & Mitchell, R. C. (1990). Determination of protein secondary structure using factor analysis of infrared spectra. *Biochemistry*, *29*(39), 9185–9193.
Li, H., Lantz, R., & Du, D. (2019). Vibrational approach to the dynamics and structure of protein amyloids. *Molecules*, *24*(1), 186.
Li, X., Mooney, P., Zheng, S., Booth, C. R., Braunfeld, M. B., Gubbens, S., Agard, D. A., & Cheng, Y. (2013). Electron counting and beam-induced motion correction enable near-atomic-resolution single-particle cryo-EM. *Nature Methods*, *10*(6), 584.
Lin, J. C., Glover, Z. K., & Sreedhara, A. (2015). Assessing the utility of circular dichroism and FTIR spectroscopy in monoclonal-antibody comparability studies. *Journal of Pharmaceutical Sciences*, *104*(12), 4459–4466.
Lin-Vien, D., Colthup, N. B., Fateley, W. G., & Grasselli, J. G. (1991). *The handbook of infrared and Raman characteristic frequencies of organic molecules*. Elsevier.
Liu, D., Ren, D., Huang, H., Dankberg, J., Rosenfeld, R., Cocco, M. J., Li, L., Brems, D. N., & Remmele, R. L., Jr (2008). Structure and stability changes of human IgG1 Fc as a consequence of methionine oxidation. *Biochemistry*, *47*(18), 5088–5100.

Liu, H., Gaza-Bulseco, G., Chumsae, C., & Newby-Kew, A. (2007). Characterization of lower molecular weight artifact bands of recombinant monoclonal IgG1 antibodies on non-reducing SDS-PAGE. *Biotechnology Letters*, *29*(11), 1611–1622.

Liu, L., Braun, L. J., Wang, W., Randolph, T. W., & Carpenter, J. F. (2014). Freezing-induced perturbation of tertiary structure of a monoclonal antibody. *Journal of Pharmaceutical Sciences*, *103*(7), 1979–1986.

Maggio, R. M., Calvo, N. L., Vignaduzzo, S. E., & Kaufman, T. S. (2014). Pharmaceutical impurities and degradation products: Uses and applications of NMR techniques. *Journal of Pharmaceutical and Biomedical Analysis*, *101*, 102–122.

Maulik, V. T., Jennifer, S. L., & Teruna, J. S. (2009). The role of thiols and disulfides in protein chemical and physical stability. *Current Protein and Peptide Science*, *10*, 614–625.

Mcauley, A., Jacob, J., Kolvenbach, C. G., Westland, K., Lee, H. J., Brych, S. R., Rehder, D., Kleemann, G. R., Brems, D. N., & Matsumura, M. (2008). Contributions of a disulfide bond to the structure, stability, and dimerization of human IgG1 antibody CH3 domain. *Protein Science*, *17*(1), 95–106.

McCreery, R. L. (2005). *Raman spectroscopy for chemical analysis* (Vol. 225). John Wiley & Sons.

Miller, L. M., Bourassa, M. W., & Smith, R. J. (2013). FTIR spectroscopic imaging of protein aggregation in living cells. *Biochimica et Biophysica Acta (BBA)-Biomembranes*, *1828*(10), 2339–2346.

Murphy, R. M. (1997). Static and dynamic light scattering of biological macromolecules: What can we learn? *Current Opinion in Biotechnology*, *8*(1), 25–30.

Ouellette, D., Alessandri, L., Chin, A., Grinnell, C., Tarcsa, E., Radziejewski, C., & Correia, I. (2010). Studies in serum support rapid formation of disulfide bond between unpaired cysteine residues in the VH domain of an immunoglobulin G1 molecule. *Analytical Biochemistry*, *397*(1), 37–47.

Pandya, A., Howard, M. J., Zloh, M., & Dalby, P. A. (2018). An evaluation of the potential of nmr spectroscopy and computational modelling methods to inform biopharmaceutical formulations. *Pharmaceutics*, *10*(4), 165.

Paterson, Y., Englander, S. W., & Roder, H. (1990). An antibody binding site on cytochrome c defined by hydrogen exchange and two-dimensional NMR. *Science*, *249*(4970), 755–759.

Pavia, D. L., Lampman, G. M., Kriz, G. S., & Vyvyan, J. A. (2008). *Introduction to spectroscopy*. Cengage Learning.

Pellecchia, M., Sem, D. S., & Wüthrich, K. (2002). NMR in drug discovery. *Nature Reviews. Drug Discovery*, *1*(3), 211–219.

Pelletier, M. J. (2003). Quantitative analysis using Raman spectrometry. *Applied Spectroscopy*, *57*(1), 20A–42A.

Poppe, L., Jordan, J. B., Lawson, K., Jerums, M., Apostol, I., & Schnier, P. D. (2013). Profiling formulated monoclonal antibodies by 1H NMR spectroscopy. *Analytical Chemistry*, *85*(20), 9623–9629.

Reddy, P. T., Brinson, R. G., Hoopes, J. T., McClung, C., Ke, N., Kashi, L., Berkmen, M., & Kelman, Z. (2018). Platform development for expression and purification of stable isotope labeled monoclonal antibodies in Escherichia coli. In *mAbs* (Vol. 10, pp. 992–1002). Taylor & Francis, No. 7.

Reichert, J.M., (2020). *The Antibody Society*. Antibody therapeutics approved or in regulatory review in the EU or US https://www.antibodysociety.org/resources/approved-antibodies/ Accessed 27 Mar 2020.

Remmele, R. L., Jr, Callahan, W. J., Krishnan, S., Zhou, L., Bondarenko, P. V., Nichols, A. C., Kleemann, G. R., Pipes, G. D., Park, S., Fodor, S., & Kras, E. (2006). Active dimer of Epratuzumab provides insight into the complex nature of an antibody aggregate. *Journal of Pharmaceutical Sciences*, *95*(1), 126–145.

Riddles, P. W., Blakeley, R. L., & Zerner, B. (1983). [8] Reassessment of Ellman's reagent. In *Methods in enzymology* (Vol. 91, pp. 49–60). Academic Press.

Robinson, P. J., Pringle, M. A., Woolhead, C. A., & Bulleid, N. J. (2017). Folding of a single domain protein entering the endoplasmic reticulum precedes disulfide formation. *Journal of Biological Chemistry, 292*(17), 6978–6986.

Rodríguez-Mendieta, I. (2015). Biophysical analysis. *BioProcess International, 13*, 8.

Rovnyak, D., Hoch, J. C., Stern, A. S., & Wagner, G. (2004). Resolution and sensitivity of high field nuclear magnetic resonance spectroscopy. *Journal of Biomolecular NMR, 30* (1), 1–10.

Rudikoff, S., & Pumphrey, J. G. (1986). Functional antibody lacking a variable-region disulfide bridge. *Proceedings of the National Academy of Sciences, 83*(20), 7875–7878.

Rusinova, E., Tretyachenko-Ladokhina, V., Vele, O. E., Senear, D. F., & Ross, J. A. (2002). Alexa and Oregon Green dyes as fluorescence anisotropy probes for measuring protein—-protein and protein—nucleic acid interactions. *Analytical Biochemistry, 308*(1), 18–25.

Rygula, A., Majzner, K., Marzec, K. M., Kaczor, A., Pilarczyk, M., & Baranska, M. (2013). Raman spectroscopy of proteins: A review. *Journal of Raman Spectroscopy, 44* (8), 1061–1076.

Saphire, E. O., Parren, P. W., Pantophlet, R., Zwick, M. B., Morris, G. M., Rudd, P. M., Dwek, R. A., Stanfield, R. L., Burton, D. R., & Wilson, I. A. (2001). Crystal structure of a neutralizing human IGG against HIV-1: A template for vaccine design. *Science, 293*(5532), 1155–1159.

Sauer, M., Hofkens, J., & Enderlein, J. (2011). Basic principles of fluorescence spectroscopy. *Handbook of fluorescence spectroscopy and imaging: from single molecules to ensembles*, 1–30.

Scapin, G., Yang, X., Prosise, W. W., McCoy, M., Reichert, P., Johnston, J. M., Kashi, R. S., & Strickland, C. (2015). Structure of full-length human anti-PD1 therapeutic IgG4 antibody pembrolizumab. *Nature Structural & Molecular Biology, 22*(12), 953.

Schauenstein, E. S., Dachs, F., Reiter, M., Gombotz, H., & List, W. (1986). Labile disulfide bonds and free thiol groups in human IgG I. *International Archives of Allergy and Immunology, 80*(2), 174–179.

Scheiblhofer, S., Laimer, J., Machado, Y., Weiss, R., & Thalhamer, J. (2017). Influence of protein fold stability on immunogenicity and its implications for vaccine design. *Expert Review of Vaccines, 16*(5), 479–489.

Singh, P., Panchaud, A., & Goodlett, D. R. (2010). Chemical cross-linking and mass spectrometry as a low-resolution protein structure determination technique. *Analytical Chemistry, 82*(7), 2636–2642.

Smith, B. C. (2011). *Fundamentals of Fourier transform infrared spectroscopy*. CRC press.

Smyth, M. S., & Martin, J. H. J. (2000). X-ray crystallography. *Molecular Pathology, 53*(1), 8.

Spiro, T. G., & Gaber, B. P. (1977). Laser Raman scattering as a probe of protein structure. *Annual Review of Biochemistry, 46*(1), 553–570.

Surewicz, W. K., Szabo, A. G., & Mantsch, H. H. (1987). Conformational properties of azurin in solution as determined from resolution-enhanced Fourier-transform infrared spectra. *European Journal of Biochemistry, 167*(3), 519–523.

Susi, H., & Byler, D. M. (1986). [13] Resolution-enhanced fourier transform infrared spectroscopy of enzymes. *Methods in Enzymology, 130*, 290–311.

Susi, H., Timasheff, S. N., & Stevens, L. (1967). Infrared spectra and protein conformations in aqueous solutions I. The amide I band in H2O and D2O solutions. *Journal of Biological Chemistry, 242*(23), 5460–5466.

Teale, F. W. J., & Weber, G. (1957). Ultraviolet fluorescence of the aromatic amino acids. *Biochemical Journal, 65*(3), 476.

Tu, A.T., & Tu, A.T., (1982). Raman spectroscopy in biology: principles and applications.

Usoltsev, D., Sitnikova, V., Kajava, A., & Uspenskaya, M. (2019). Systematic FTIR spectroscopy study of the secondary structure changes in human serum albumin under various denaturation conditions. *Biomolecules, 9*(8), 359.

Van Buren, N., Rehder, D., Gadgil, H., Matsumura, M., & Jacob, J. (2009). Elucidation of two major aggregation pathways in an IgG2 antibody. *Journal of Pharmaceutical Sciences*, *98*(9), 3013–3030.

Visser, J., Feuerstein, I., Stangler, T., Schmiederer, T., Fritsch, C., & Schiestl, M. (2013). Physicochemical and functional comparability between the proposed biosimilar rituximab GP2013 and originator rituximab. *BioDrugs*, *27*(5), 495–507.

Vivian, J. T., & Callis, P. R. (2001). Mechanisms of tryptophan fluorescence shifts in proteins. *Biophysical Journal*, *80*(5), 2093–2109.

Wang, H. W., & Wang, J. W. (2017). How cryo-electron microscopy and X-ray crystallography complement each other. *Protein Science*, *26*(1), 32–39.

Wang, X., Li, Q., & Davies, M. (2014). Higher-order structure comparability. *BioProcess International*, *12*, 6.

Waxdal, M. J., Konigsberg, W. H., Henley, W. L., & Edelman, G. M. (1968). The covalent structure of a human γG-immunoglobulin. II. Isolation and characterization of the cyanogen bromide fragments. *Biochemistry*, 7(5), 1959–1966.

Wei, Z., Shacter, E., Schenerman, M., Dougherty, J., & McLeod, L. D. (2011). The role of higher-order structure in defining biopharmaceutical quality. *BioProcess International*, *9*(4), 58.

Wen, Z. Q. (2007). Raman spectroscopy of protein pharmaceuticals. *Journal of Pharmaceutical Sciences*, *96*(11), 2861–2878.

Wen, Z. Q., & Thomas, G. J., Jr (1998). UV resonance Raman spectroscopy of DNA and protein constituents of viruses: assignments and cross sections for excitations at 257, 244, 238, and 229 nm. *Biopolymers: Original Research on Biomolecules*, *45*(3), 247–256.

Williams, R. W. (1986). [14] Protein secondary structure analysis using Raman amide I and amide III spectra. In *Methods in enzymology* (Vol. 130, pp. 311–331). Academic press.

Williamson, R. T., & Marquez, B. L. (2017). Small molecule NMR in pharmaceutical sciences. *Magnetic Resonance in Chemistry*, *55*(4), 252-252.

Winther, J. R., & Thorpe, C. (2014). Quantification of thiols and disulfides. *Biochimica et Biophysica Acta (BBA)-General Subjects*, *1840*(2), 838–846.

Xiang, T., Chumsae, C., & Liu, H. (2009). Localization and quantitation of free sulfhydryl in recombinant monoclonal antibodies by differential labeling with 12C and 13C iodoacetic acid and LC–MS analysis. *Analytical Chemistry*, *81*(19), 8101–8108.

Yang, H., Yang, S., Kong, J., Dong, A., & Yu, S. (2015). Obtaining information about protein secondary structures in aqueous solution using Fourier transform IR spectroscopy. *Nature Protocols*, *10*(3), 382.

Yao, J., & Yoon, J. (2000). Low-noise electrometer and its low-noise cryogenic probe with completely guarded sample chamber. *Review of Scientific Instruments*, *71*(4), 1776–1780.

Zhang, T., Zhang, J., Hewitt, D., Tran, B., Gao, X., Qiu, Z. J., Tejada, M., Gazzano-Santoro, H., & Kao, Y. H. (2012). Identification and characterization of buried unpaired cysteines in a recombinant monoclonal IgG1 antibody. *Analytical Chemistry*, *84*(16), 7112–7123.

Zhang, W., & Czupryn, M. J. (2002). Free sulfhydryl in recombinant monoclonal antibodies. *Biotechnology Progress*, *18*(3), 509–513.

Zhang, W., Marzilli, L. A., Rouse, J. C., & Czupryn, M. J. (2002). Complete disulfide bond assignment of a recombinant immunoglobulin G4 monoclonal antibody. *Analytical Biochemistry*, *311*(1), 1–9.

Zheng, K., Bantog, C., & Bayer, R. (2011). The impact of glycosylation on monoclonal antibody conformation and stability. In *MAbs* (Vol. 3, pp. 568–576). Taylor & Francis, No. 6.

Zheng, K., Middaugh, C. R., & Siahaan, T. J. (2009). Evaluation of the physical stability of the EC5 domain of E-cadherin: Effects of pH, temperature, ionic strength, and disulfide bonds. *Journal of Pharmaceutical Sciences*, *98*(1), 63–73.

CHAPTER 7

Advanced structural elucidation of new antibody formats by mass spectrometry

Tobias Graf, Michael Leiss, Markus Haberger and Patrick Bulau
Pharma Technical Development, Roche Diagnostics GmbH, Penzberg, Germany

7.1 Introduction

Recent breakthroughs in the treatment of various severe diseases are irrefutably intertwined with the emergence of protein-based biopharmaceuticals (Evens & Kaitin, 2015). Boasting the approval of numerous medications in the last decades, such as recombinant signaling proteins, monoclonal antibodies (mAbs), and fusion proteins, biotechnology ascended from a niche market to a vital and prosperous pillar within the pharmaceutical industry (Evens, 2016; Kesik-Brodacka, 2018). From the beginning, mAbs occupied a pivotal role in this rapidly evolving field and have retained their position up to today, accounting for 60% of total biopharmaceutical approvals between 2016 and 2019 in the United States (FDA, 2020). This unprecedented and continuing success story is closely linked to the unique pharmacological properties of mAbs, including their high target specificity, long serum half-life, and selectively variable effector functions (Chames, Van Regenmortel, Weiss, & Baty, 2009).

Yet despite the advantages over conventional small-molecule drugs, protein-based bio-therapeutics present a number of analytical challenges. Originating from their large molecular size and production within living organisms, natural variability at the proteome level is an inherent feature of mAbs and other biologics (Vulto & Jaquez, 2017). The product complexity is augmented by structural and conformational dynamics (Houde & Engen, 2013; Majumdar et al., 2015), a multitude of different process- and product-related impurities, as well as the characteristic microheterogeneity arising through posttranslational modifications (PTMs) (Beck & Liu, 2019; Jefferis, 2016; Jenkins, Murphy, & Tyther, 2008), sequence variants (Borisov, Alvarez, Carroll, & Brown, 2015; Yang et al., 2010), and

Monoclonal Antibodies
DOI: https://doi.org/10.1016/B978-0-12-822318-5.00005-3
Copyright © 2021 Elsevier Inc.
All rights reserved.

partial chemical modification/degradation of certain susceptible amino acids (Grassi & Cabrele, 2019; Schmid et al., 2018). To cope with these difficulties, comprehensive characterization, consistent quality testing, and appropriate control strategies play a fundamental role in confirming the drug's composition and ensuring its efficacy, purity, stability, and, most critically, safety for patients (Robinson & Jones, 2011). Based on the outlined aspects, so-called critical quality attributes (CQAs) need to be identified and monitored during early development and the target product profile progressively refined with advancing product-/process knowledge (Alt et al., 2016; Singh, Kumar, & Rathore, 2017). To characterize and monitor these, a wide array of different bioanalytical, biophysical, and immunological techniques have been developed and implemented over the last decades and are now routine apparatus in a product's life-cycle (Robinson & Jones, 2011). Among the methods for routine testing in quality control (QC) laboratories, liquid chromatography (LC)- and capillary electrophoresis (CE)-based techniques coupled with ultraviolet/fluorescence detection (UV/FLD) detection are usually the methods of choice to track product-specific CQAs (e.g., size-, charge- and glycan variants), while mass spectrometry (MS) is mainly performed during comparability exercises and for characterization of physicochemical methods (Robinson & Jones, 2011). The former, fulfilling the highest demands in terms of simplicity, ease of use and robustness, are ideally suited for routine use in stringently regulated QC environments following International Council for Harmonization (ICH) and good manufacturing practice (GMP) standards. Up until now, these established methods have repeatedly demonstrated their suitability for quality testing of antibody products. However, in view of a new, emerging generation of increasingly complex biopharmaceuticals, it is questionable as to whether these analytical tools will suffice to meet the growing requirements for future quality testing regimes.

Technical advances in protein engineering, accumulation of knowledge and hands-on experience as well as the urgent need for innovative treatment modalities have propelled the development of numerous alternatives to standard immunoglobulin G (IgG)-type formats. For instance, the class of bispecific antibodies (bsAbs), conceptualized to recognize two different target epitopes simultaneously, gave rise to a broad spectrum of commercialized technology platforms for the generation of bivalent scaffolds ranging from antigen-binding fragments to IgG-like molecules with additional domains attached (Klein, Schaefer, & Regula, 2016; Labrijn, Janmaat, Reichert, & Parren, 2019; Suurs, Lub-de Hooge, de Vries, & de Groot,

2019). In addition to highly elaborate protein designs, new treatment options such as antibody–drug conjugates (ADCs) - (Birrer, Moore, Betella, & Bates, 2019; Ehkirch, Goyon, et al., 2018) bio-conjugates consisting of mAbs interconnected with cytotoxic agents by a chemical linker and antibody coformulations (Cao et al., 2019) - have introduced an additional dimension of complexity. The immense potential of these new modalities is reflected in the rising number of such drugs already approved or evaluated in clinical studies. In Table 7.1 the currently marketed examples are summarized, including two bsAbs, eight ADCs, and two coformulations (EMA, 2020; FDA, 2020). Furthermore, more than 250 products belonging to these categories are presently undergoing phase I, II, or III clinical trials (NIH, 2020). Analytical challenges hence faced include new size variants derived from mispairing of varying light and heavy chains, correct assignment of PTMs to a given subdomain species and, in general, a higher degree of heterogeneity resulting in unresolvable peak patterns. To address the complexity, higher resolution methods are imperative, for which LC–MS offers great promise. In this context, multiattribute method (MAM) approaches are receiving significant attention. At present, however,

Table 7.1 Approved products in the United States/EU exceeding the molecular complexity of immunoglobulin G molecules (EMA, 2020; FDA, 2020).

Active ingredient	Brand name
ADCs	
Brentuximab vedotin	ADCETRIS
Enfortumab vedotin-ejfv	PADCEV
Gemtuzumab ozogamicin	MYLOTARG
Polatuzumab vedotin-piiq	POLIVY
Fam-trastuzumab deruxtecan	ENHERTU
Trastuzumab emtansine	KADCYCLA
Sacituzumab govitecan	TRODELVY
Inotuzumab ozogamicin	BESPONSA
Bispecific antibodies	
Blinatumomab	BLINCYTO
Catumaxomab[a]	Removab
Emicizumab-kxwh	Hemlibra
Coformulations	
Trastuzumab/hyaluronidase-oysk	Herceptin Hylecta
Trastuzumab, pertuzumab, and hyaluronidase-zzfx	PHESGO

ADC, Antibody–drug conjugate.
[a]Withdrawn from the market for commercial reasons.

established MAM methodologies are mostly restricted to bottom-up LC—MS applications that can only provide information at (reduced) peptide level and are hence incapable of describing the proteoform distribution in the native protein state (Rogers et al., 2017).

These constraints of conventional protein characterization have motivated the use of complementary methods to facilitate identification, quantification, and monitoring of CQAs while preserving the molecule's integrity. For this purpose, coupling the versatility and robustness of LC approaches with the selectivity and sensitivity of MS has proven highly useful in a vast number of recent studies, a review of which will be presented here. In particular, with the increasing commercial availability of two-dimensional (2D) LC systems and high-resolution and mass range mass spectrometers, as well as major technical advancements made in both fields within the last decade, comprehensive characterization of new biotherapeutic formats at native protein level has now become a tangible objective.

7.2 Product attributes and the applied methods for their characterization

Product-related impurities constitute a major class of potential CQAs in biopharmaceuticals that are, as defined in ICH guideline Q6B (ICH, 1999), "molecular variants arising during manufacture and/or storage, which do not have properties comparable to those of the desired product with respect to activity, efficacy, and safety." These involve, for instance, a vast number of posttranslational and chemical-induced modifications originating from their production in living organisms and their susceptibility to degradation pathways during processing and storage. Based on the molecular properties relative to the desired product, they can be subdivided into the following classes: size (1), charge (2), oxidation (3), glycosylation (4), and structural and sequence variants (5). For characterization and monitoring of these product attributes, suitable analytical methods need to be developed and applied, which can, in general, be assigned to two categories: release testing and stability indicating methods (1) and methods for (extended) characterization (2). Common product attributes belonging to the five categories as well as the applied techniques for release/stability testing and extended characterization, as exemplified for mAbs, are further discussed below and summarized in Table 7.2.

Table 7.2 List of common product attributes for monoclonal antibodies.

Attribute category	Product attribute	Analytical methods
(1) Size variants	High–molecular weight variants (HMWs) Low–molecular weight variants (LMWs)	Release: SE-(U)HPLC, CE Extended: SEC-MALS, AUC
(2) Charge variants	Asn deamidation	Release: IEC, cIEF; Extended: LC–MS peptide mapping
	Asp isomerization Asu N-terminal variants C-terminal variants Lys glycation	
(3) Oxidation variants	Met/Trp/Cys oxidation	Release: HIC Extended: LC–MS peptide mapping
(4) Glycosylation variants	Fc and Fab glycosylation	Release: HILIC–FLD/MS or CGE–LIF of labeled glycans Extended: LC–MS peptide mapping
(5) Structural and sequence variants	As a consequence of (1)–(4)	See above
	Free thiol	Release: Spectrophotometric assays Extended: LC–MS peptide mapping
	Sequence variants	Release: LC–UV peptide mapping Extended: LC–MS peptide mapping
	Higher order structure	Extended: CD, intrinsic fluorescence, Raman and FT-IR spectroscopy

Asu, Aspartyl succinimide; *AUC*, analytical ultracentrifugation; *CD*, circular dichroism; *CE*, capillary electrophoresis; *CGE*, capillary-gel electrophoresis; *cIEF*, capillary isoelectric focusing; *FLD*, fluorescence detection; *FT*-IR, Fourier-transform infrared; *HIC*, hydrophobic interactions chromatography; *HILIC*, hydrophilic interaction chromatography; *LC*, liquid chromatography; *LIF*, laser-induced fluorescence; *MALS*, multiangle light scattering; *MS*, mass spectrometry; *UV*, ultraviolet.

See also Chapter 2, Physicochemical Characterization of Monoclonal Antibodies, for biophysical methods employed for standard monoclonal methods.

1. The main concern with respect to size variants in conventional biotherapeutics is the potential immunogenicity of high–molecular weight variants (HMWs, e.g., dimers, trimers, and higher order

aggregates), which may induce severe hypersensitivity responses (Moussa et al., 2016; Rosenberg, 2006) and potentially affect product efficacy (Roberts, 2014) due to the induction of antidrug antibodies. In addition, low–molecular weight variants (LMWs, e.g., fragments) are often associated with the loss of biological activity and unfavorable pharmacokinetics. For the assessment of size heterogeneity, size exclusion (SE) chromatography [combined with high-performance LC, SE-HPLC, or ultra high-performance LC, SE-UHPLC (Lowe et al., 2011)] and/or CE (Rustandi, Washabaugh, & Wang, 2008) using UV detection are generally the methods of choice for routine product testing and characterization. Other conventional approaches to obtain molecular mass and size distribution information include coupling of the chromatography system with a multiangle light scattering (Ye, 2006) detector or application of analytical ultracentrifugation (Lowe et al., 2011).

2. Capillary isoelectric focusing electrophoresis and ion exchange HPLC (IE-HPLC) are the most common methods for monitoring charge-related variants (Vlasak & Ionescu, 2008). Based on the elution behavior relatively to the product target (usually representing the main peak), this class of process-related impurities can be separated into acidic and basic charge variants. The former commonly involves deamidation of asparagine, sialylation, and glycation of lysine residues, while oxidation, removal of sialic acid, glycine amidation, N-terminal formation of pyroglutamic acid, and incomplete removal of C-terminal lysine residues typically constitute characteristic product peaks in the basic region (Beck, Wagner-Rousset, Ayoub, Van Dorsselaer, & Sanglier-Cianferani, 2013). Isomerization of aspartate residues to isoaspartic acid (iso-Asp) and its reaction intermediate succinimide are also frequently occurring acidic or basic species (Zhang, Yip, & Katta, 2011). In general, chemical and PTMs in the complementary-determining region (CDR) of IgGs are likely to lead to an alteration of target binding activity and are therefore considered (potential) CQAs (Haberger et al., 2014). As the abovementioned profile-based methods only provide a low-resolution representation of the charge variant distribution, the exact positions need to be additionally determined by site mapping using LC–MS after proteolytic digestion (Dick, Mahon, Qiu, & Cheng, 2009).
3. Protein oxidation is a chemical modification, which is induced by reactive oxygen species, and can be observed predominantly on

methionine and, to a lesser extent, tryptophan and cysteine residues (Folzer et al., 2015; Loew et al., 2012). Oxidation of the conserved methionine residues at the solvent-exposed CH2–CH3 interface of recombinant IgG1 and IgG2 antibodies reportedly affects their affinity to the neonatal Fc receptor and shortens circulation half-life (Bertolotti-Ciarlet et al., 2009; Pan et al., 2009; Stracke et al., 2014). Moreover, oxidation-induced conformational changes in the CDRs have a potential impact on target binding (Wei et al., 2007). Based on the chromatographic behavior of oxidized species, hydrophobic interactions chromatography (HIC) is the standard method applied for the analysis of these modifications in QC (Boyd, Kaschak, & Yan, 2011; Haverick, Mengisen, Shameem, & Ambrogelly, 2014). Analogous to charge-related variants, space-resolved assignment of protein oxidation at amino acid level is conducted by peptide mapping using LC–MS techniques (Beck et al., 2013).

4. Many bio-therapeutics, including mAbs, hormones, and fusion proteins, feature N-/O-glycans. In contrast to the well-defined primary amino acid sequence of recombinant proteins, glycosylation is subject to inherent variability resulting in a wide spectrum of heterogeneous glycoforms with a potential impact on stability, pharmacokinetics and pharmacodynamics, bioactivity, and safety (Chung et al., 2008; Higel, Seidl, Sorgel, & Friess, 2016; Jennewein & Alter, 2017). Among them, attachment of fucose, sialic acid, and galactose to the N-linked glycan core as well as nonglycosylated and high-mannose type glycans need to be controlled and monitored throughout development, processing, and manufacturing of glyco-proteins (del Val, Kontoravdi, & Nagy, 2010). Evaluation of glycosylation as a quality attribute requires the determination of glycan structures and their relative quantity as well as identification of the glycosylation site(s) on the protein (Beck et al., 2013; de Haan, Falck, & Wuhrer, 2019). The former tasks are usually performed by derivatizing glycans after cleavage with commercially available labeling kits [e.g., 2-aminobenzoic acid (2-AA), 2-aminobenzamide (2-AB), 8-aminopyrene-1,3,6-trisulfonic acid (APTS), and GlycoWorks RapiFluor-MS] followed by hydrophilic interaction chromatography (HILIC) or capillary-gel electrophoresis with laser-induced fluorescence analysis (Reusch et al., 2014; Reusch, Haberger, Falck, et al., 2015). Bottom-up approaches after proteolytic digestion using LC coupled to tandem MS (MS/MS) facilitate the site- and glycan-specific analysis (Reusch, Haberger, Maier, et al., 2015).

See also Chapter 3, Glycosylation Analysis.

5. The primary structure of proteins (amino acid sequence), PTMs, chemical modifications, and the location of disulfide bonds dictate the higher order structure of proteins (i.e., secondary, tertiary, and quaternary structures). Confirmation of the primary structure is hence fundamental to the structural characterization of proteins. The presence of sequence variants therefore needs to be controlled, especially as they have a potential yet difficult-to-predict effect on the properties, function, and immunogenicity of recombinant proteins (Yang et al., 2010). The number and positions of intra- and interchain disulfide linkages also play a crucial role in stabilizing the correct structure of mAbs and other multichain proteins, critical to maintaining their biological function (Hagihara & Saerens, 2014). In this context, incomplete formation (i.e., presence of free thiols) of disulfide bonds may affect, for instance, bioactivity, pharmacokinetics, and immunogenicity of the product (Eon-Duval, Broly, & Gleixner, 2012; Harris, 2005; Li, Haque, & Blum, 2002).

   To examine these quality attributes, spectroscopy-based methods (in the case of free sulfhydryls) and LC–MS peptide mapping are the routinely used methods (Borisov et al., 2015; Lakbub, Shipman, & Desaire, 2018; Xiang, Chumsae, & Liu, 2009). In addition to the analysis of the covalent arrangement of atoms within a bio-therapeutic, the importance of assessing the higher order structure to confirm the correct native state, whereby assuring full biological impact and reduced risk of potential side-effects, is becoming increasingly recognized (Gabrielson & Weiss, 2015; Huang & Chen, 2014; Maas, Hermeling, Bouma, Jiskoot, & Gebbink, 2007). Common biophysical methods to detect drug product variants occurring during drug manufacturing include (vibrational) circular dichroism (Li et al., 2011; Lin, Glover, & Sreedhara, 2015; Nagarkar, Murphy, Yu, Manning, & Al-Azzam, 2013), Raman (Kitagawa, 1992), intrinsic fluorescence (Blake-Haskins, Lin, Wu, Perkins, & Spitznagel, 2014), and Fourier-transform infrared spectroscopy (Lin et al., 2015). These techniques provide characteristic fingerprints based on the protein conformation, but only allow the elucidation of the higher order structure with low resolution and are thus less suitable to observe small conformational variances and structural dynamics. Potential alternatives to assess the structure–function relationship of CQAs in more detail, such as 2D-nuclear magnetic resonance and MS-based methods (e.g., hydrogen-deuterium exchange, covalent protein footprinting and crosslinking MS) are therefore becoming increasingly adopted. For an elaborate portrayal of emerging technologies to determine the higher order structure of bio-therapeutics, the reader is referred to recent publications covering this topic (Auclair, Rathore, & Krull, 2019; Bonnington et al., 2017; Brinson, Arbogast, Marino, & Delaglio, 2020; Marino et al., 2015; Misra, 2017).

Over the last decades, identifying and monitoring (potential) CQAs of standard biopharmaceutical formats remained a tedious but relatively straightforward exercise using the delineated analytical toolbox. For instance, the analysis of trastuzumab involves an array of different GMP-compliant techniques capable of covering all required aspects for QC testing. Assessment of size variants (e.g., dimers, trimers, and hinge-region fragmentation products) can be adequately performed by a combination of SEC and CE- sodium dodecyl sulfate (SDS) (Kim et al., 2017; Schmid et al., 2018). Likewise, charge variants, which have a potential impact on product efficacy (e.g., HC-iso-Asp 102, HC-iso-Asp 283, HC-iso-Asp

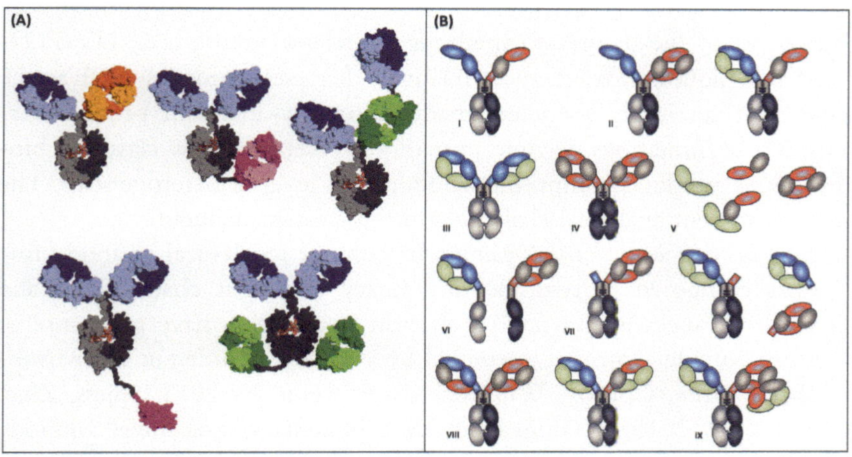

Figure 7.1 Schematic illustration of complex antibody formats and their by-products. (A) Recent advances in antibody design and manufacturing have led to a vast array of antibody scaffolds featuring great structural and functional variety as illustrated by a nonexhaustive selection of complex antibody formats. (B) Compared to standard IgG-like molecules, the number of theoretically occurring by-products is considerably increased as demonstrated for a bispecific 1 + 1 CrossMAb format: (I) HC dimer; (II) 3/4 MAb without LC1 or LC2; (III) knob—knob dimer; (IV) hole—hole dimer; (V) free LC and LC Homo- and Heterodimers; (VI) half MAbs; (VII) Fc-Fab MAbs and Fabs; (VIII) mispaired LCs; (IX) CrossMAb + LC-Heterodimer. *LC*, Liquid chromatography.

404, LC-Asn30-deam, HC-Asn55-deam and HC-Lys65-glycation), are properly resolved by IEC (Bailey et al., 2018; Haberger et al., 2014; Schmid et al., 2018). Initial peak assignment is generally achieved by off-line fractionation in preparative scale followed by variant site mapping employing bottom-up LC−MS analysis. Furthermore, potential hot-spots (amino acids susceptible to modifications) are identified by conducting forced stress studies (e.g., elevated temperature, glucose, and oxidation conditions) (Diepold et al., 2012; Wei, Berning, Quan, & Zhang, 2017). It could be demonstrated that the standard methods applied for routine testing were capable of resolving all critical variants and detecting stability-related changes over time. For evaluation of glycosylation, a 2-AB labeling approach using HILIC−FLD analysis adequately covers all relevant *N*-glycan species originating from the glycosylation site at the CH2 domain on each heavy chain, including the variants with a potential impact on product quality (e.g., Man5/6, galactosylation, afucosylation, and attached sialic acid) (Kuhne et al., 2019).

7.3 Limitations of the analytical toolbox

Drawbacks of the delineated methods for release testing (i.e., UV/FLD-based techniques) involve their relatively low resolution as well as the insufficient sensitivity to detect medium or low-abundant product variants. These limitations become primarily apparent for new classes of bio-therapeutic products comprising an additional level of heterogeneity. The most prominent analytical challenges in this context include:

1. *New variants by chain-(mis)pairing*: In contrast to classical antibody formats composed of two identical heavy and light chains (HCs and LCs), advances in the field of protein engineering have given rise to more complex formats, assembled by a variety of different chains/subunits (Byrne, Conroy, Whisstock, & O'Kennedy, 2013; Spiess, Zhai, & Carter, 2015). A nonexhaustive representation of novel antibody structures is depicted in Fig. 7.1A. Recombinant production of these molecules is often accompanied with a higher number of product-related by-products and aggregates resulting from homo-dimerization, heavy chain−light chain mispairing, and intermolecular interactions (Fan, Wang, Hao, & Li, 2015; Haberger, Leiss, et al., 2016; Klein et al., 2016), as shown exemplarily for a 1 + 1 CrossMAb format in Fig. 7.1B. Achieving resolution and hence characterization of the resulting heterogenic mix is not always trivial, as in some instances

these by-products feature similar molecular properties (based on their size, charge variant distribution or hydrophobicity) compared to the desired product or other product-related impurities and degradants present (Macchi et al., 2015; Woods, Xie, Von Kreudenstein, Ng, & Dixit, 2013). Due to this fact, identification and quantification of these species at native protein level are challenging when using conventional UV-based chromatographic approaches.

2. *Increased micro-heterogeneity*: Incorporation of additional building blocks [e.g., drug payload of ADCs (Wagh, Song, Zeng, Tao, & Das, 2018), extra subunits attached via linkers for Fusion proteins (Valedkarimi, Nasiri, Aghebati-Maleki, & Majidi, 2017)] into the molecule's scaffold and modifications associated with their characteristic features [e.g., O-glycosylation, phosphorylation, and presence of hydroxyproline in glycine-serine linkers (Spahr, Gunasekaran, Walker, & Shi, 2017; Spahr, Shi, & Lu, 2014; Tyshchuk et al., 2017)] introduce a further dimension of variability at the proteoform level. Improved fermentation processes that are designed to increase titers of recombinant biopharmaceuticals can also potentially increase complexity in glycosylation patterns due to uncompleted glycan processing (Karst, Steinebach, & Morbidelli, 2018). In such cases, profile-based LC methods with only UV detection can prove inadequate to resolve the resulting complex proteoform distributions (e.g., due to overlap of multiple different modifications).
3. *Detection of low-abundant species*: The emergence of highly potent products (e.g., immune-modulators, ADCs, Cytokine-fusion proteins) requires an improved sensitivity to detect low-abundant species potentially impacting product safety, which is particularly true for the chain-pairing variants described above (Lee et al., 2019). Frequently, the conventional toolbox does not offer sufficient sensitivity or resolution to quantify trace-amounts of critical species.
4. *Spatial distribution of PTMs*: The complexity of recent formats is further complicated by an increased number of different building blocks, which, due to their derivation from the ancestral IgG backbone, share a high degree of sequence homology amongst each other (Gunasekaran et al., 2010). This sequence homology restricts the capability of common bottom-up peptide mapping approaches to assign PTMs to specific domains in the molecule and underscores the need for analytical approaches preserving the quaternary structure of biotherapeutics.

In light of these challenges facing the analysis of next-generation biotherapeutics, conventional approaches for quality testing clearly prove inadequate to cover all product-related quality attributes. Adjustments by means of complementary analytical separation and higher resolution detection techniques, particularly by leveraging orthogonal chromatographic separation methods and MS detection are pertinent. A detailed description of the increasingly important technologies for unambiguous product characterization is presented in the following two subchapters.

7.4 The emerging role of mass spectrometry

MS has evolved into an indispensable tool for the characterization of biotherapeutics over the last decades. A data-mining study, conducted by authors from the FDA based on approved Biologics License Applications (BLAs) between 2000 and 2015, stated that MS was employed in the vast majority of all submissions and the mean number of quality attributes analyzed by MS increased steadily from 2 to 11 over the same period (Rogstad et al., 2017). The most common MS applications include amino acid sequence and product mass determination, and identification of the position and nature of PTMs or glycan structures (Rogstad et al., 2017). The expanding and diversifying role of MS for biopharmaceutical analysis primarily stems from technical advances of modern mass analyzers in terms of sensitivity, resolution, and mass accuracy. High-resolution MS is nowadays routinely available in many biopharmaceutical laboratories promoted by the launch of quadropole time-of-flight and Orbitrap mass analyzers (Wang & Bennett, 2017). Improved instrument robustness for LC–MS technologies using electrospray ionization (ESI) and the introduction of user-friendly software/hardware tailored for non-MS experts are paving the way for additional application fields for monitoring of product and process attributes during the entire manufacturing and QC process (Haberger, Bonnington, et al., 2016). Moreover, alternative fragmentation techniques such as electron-capture dissociation (ECD), electron-transfer dissociation (ETD), and electron-transfer higher energy collision dissociation (EThcD), which are increasingly implemented on a variety of instrument platforms, prove particularly valuable when combined with commonly applied collisional activating techniques (including collision-induced dissociation and higher energy dissociation) (Darula, Pap, & Medzihradszky, 2019; Yu et al., 2017; Zhu, Su, Clark, Go, & Desaire,

2013). Typically, ECD, ETD, and EThcD afford peptide fragmentation at N−Cα bonds and therefore provide a reliable tool to detect selected PTMs (e.g., *O*-glycosylation and phosphorylation) that cannot be unambiguously detected by traditional collision-induced fragmentation techniques (Brodbelt, 2016; Cole, Ma, Zhang, & Xia, 2012). In general, two distinct strategies for the MS evaluation of product attributes have emerged, the first focusing on the "bottom-up" approach [i.e., at the (glyco-)peptide level following digestion] and the second aimed at higher organizational levels (i.e., subunit, intact, or native analysis, aka "middle-up" or "top-down" approaches).

To overcome the limited specificity and resolution of the conventional release testing methods, the implementation of an LC−MS peptide-mapping workflow, referred to as MAM, into regulated (GxP) laboratories has attracted increasing interest over the last years (Berger & Fredette, 2017; Rogers et al., 2017; Zhang & Guo, 2017). After initial characterization of the product and identification of modifications by MS/MS, those quality attributes defined as critical are quantified based on the MS1 precursor ions, at best via fully automated data processing. Owing to its capacity to resolve a wide spectrum of different quality attributes (e.g., charge-, oxidation-, and glyco-variants) simultaneously, this approach holds great promise within the biopharmaceutical industry to streamline the development of new products and QC testing. Comparison of MAM workflows in recent studies with conventional methods, such as IEC and HILIC glycan analysis, resulted in highly consistent data (Buettner, Maier, Bonnington, Bulau, & Reusch, 2020; Rogers et al., 2015), corroborating the potential of MAM to complement or even replace them over the long term. Furthermore, MAM may also contribute to improved process/product understanding, in line with QbD principles, by virtue of the peptide- and location-specific information it provides with respect to modifications and attached glycans. Dedicated software packages supporting new peak detection algorithms within the automated data processing enable detection of unknown impurities, which may occur during batch release testing. A currently published study summarizes the *status quo* of MAM for QC testing and provides key aspects to consider for its application (Rogstad et al., 2019).

Assessment of bio-therapeutics at the subunit or intact level has become a viable resource to gain further knowledge on the entire protein structure. Compared to conventional bottom-up workflows, these approaches are capable of elucidating molecular properties [e.g., misparing of subunits, disulfide linkages (Faid, Leblanc, Bihoreau, & Chevreux, 2018), and combination of multiple modifications (Rogstad et al., 2019)] that are otherwise removed or scrambled in peptide sample preparation and, therefore, facilitate comprehensive characterization of distinct (glyco-)proteoforms. Moreover, MS/MS approaches at the intact protein level using UV photodissociation or ECD can be also utilized to localize PTMs of amino acids within the protein (Fornelli et al., 2012; Shaw et al., 2020; Srzentic et al., 2020). In addition, minimized sample manipulation reduces the risk for inducing experimental artifacts and renders them more robust than traditional peptide mapping approaches (Mouchahoir & Schiel, 2018).

The standard workflow for intact mass analysis involves coupling reverse-phase LC (RPLC) to MS, performed at relatively high concentrations of organic solvents and elevated column temperatures through which the native structure of the analyzed protein is not maintained (Yin et al., 2016). The consequently denatured proteins exhibit relatively high charge state distributions (CSDs), which, in turn, increase the risk of spectral overlapping and, hence, limits the resolution of multiple proteoforms in heterogeneous samples (Grandori, 2003). Reduction of spectral complexity can be achieved by the analysis of large fragments or subunits generated by limited proteolysis and/or disulfide reduction (Zang, 2017). In particular, enzymatic cleavage of heavy chains below or above the hinge region prior to the analysis and, optionally, chemical reduction in IgG-like molecules facilitates the domain-specific characterization. For this purpose, antibody digestion using proteases such as IdeS (immunoglobulin-degrading enzyme of *Streptococcus pyogenes*) or Kgp (lysine specific protease from *Porphyromonas gingivalis*) has become a common tool for the middle-up analysis of bio-therapeutics (Leblanc, Ramon, Bihoreau, & Chevreux, 2017; Sokolowska, Mo, Dong, Lewis, & Hu, 2017).

Even further information can be obtained through native MS (nMS) workflows, that is, performing intact mass analysis under "near-" native or physiological-like conditions utilizing volatile salts that are compatible with ESI at low-to-moderate concentrations (Konijnenberg, Butterer, & Sobott, 2013; Leney & Heck, 2017). In contrast to the analysis under denaturing conditions, proteins are transferred from solution into the gas phase while maintaining their native folded state, thereby restricting the solvent-exposed surface area. As a consequence, the number of charges observed by MS is considerably lowered translating into a reduced CSD and spanning a wider mass-to-charge (m/z) ratio range. This potentially enhances the overall signal intensity and reduces the risk of spectral interference (Donnelly et al., 2019; Schachner et al., 2016). In addition, non-covalent interactions can be preserved enabling the analysis of protein/protein or protein/nucleic acid complexes ranging in size from a few hundreds to thousands of kDa (Bereszczak et al., 2012; Rose, Damoc, Denisov, Makarov, & Heck, 2012; Satiaputra et al., 2019). Owing to the improved resolution of modern mass analyzers also offering high mass range detection, nMS has become a viable approach for the robust characterization of recombinant proteins such as mAbs and ADCs (Beck et al., 2016; Terral, Beck, & Cianferani, 2016). Over the last years, alternative mobile phases consisting of ESI-compatible salts have been developed for different separation modes [e.g., IEC (Leblanc et al., 2017), SEC (Haberger, Leiss, et al., 2016), HIC (Chen et al., 2016), and CE (Belov et al., 2017)], which enable online coupling to nMS (LC-nMS). Compared to infusion-based nMS applications, LC-nMS promises numerous benefits. Due to the chromatographic separation of distinct proteoforms prior to the MS analysis, mass accuracy and the dynamic range are significantly improved (Donnelly et al., 2019), and the chromatographic behavior of a given species also provides an additional source of information, which can be leveraged for peak assignment.

In this context, several authors including Fussl et al. (2018) and Bailey et al. (2018) have demonstrated that MS data generated by direct hyphenation of IEC to MS utilizing volatile salt mobile phases over a pH gradient were sufficient to detect distinct glycoforms and PTMs (Leblanc, Bihoreau, & Chevreux, 2018; Muneeruddin et al., 2017; Yan, Liu, Wang, Daly, & Li, 2018). Most notably, both workflows were able to assign single deamidated variants (featuring a mass shift of +1 Da relative to the main peak) by correlating the observed masses with retention times of the respective peaks. As expected, chromatographic baseline separation and sufficient absolute abundance of the isoforms were necessary to measure these near isobaric species unambiguously. In a follow-up study (Fussl et al., 2019) the potential of LC-nMS to identify charge variants of adalimumab was evaluated using an optimized LC gradient. Interestingly, mass analysis unveiled additional species, which were unresolvable by mere chromatographic separation. Apart from its value for peak assignment, the benefits of LC-nMS in terms of sensitivity compared to UV detection were demonstrated recently by Ma, Raoufi, Bailly, Fayadat-Dilman, and Tomazela (2020). Several truncated antibody species that remained undetected in the UV trace during the charge variant analysis of different mAb products could be reliably identified in a range between 0.05% and 0.4% using MS detection, underlining its potential for detecting low-abundant species after targeted optimization of chromatographic and MS parameters. While previous studies are limited to the charge variant analysis of standard IgG molecules, even a higher number of coeluting species can be assumed for complex formats. Given the high interest in MS-compatible IE eluent systems, it is likely that further advancements (e.g., in column design for ensuring linear pH responses) will make this technique even more powerful.

Besides its application for the analysis of charge variants, LC-nMS has been adopted for the characterization of size variants owing to the superior mass accuracy and sensitivity compared to conventional UV and light scattering detectors (Haberger, Leiss, et al., 2016). Compatibility with ESI allowing direct coupling of SEC to MS is usually achieved by replacing nonvolatile, high ionic strength buffers (e.g., phosphate salts) with volatile mobile phase conditions (e.g., ammonium acetate). In this regard, the advent of state-of-the-art columns with sub-3 μm particles has proved highly valuable due to their improved inertness minimizing secondary interactions between the analyte and the stationary phase even at low salt concentrations (Goyon, Beck, et al., 2017). Furthermore, operating electrospray at decreased flow rates as commonly applied for UHPLC improves the overall ionization efficiency. The benefits of this approach were demonstrated recently for the analysis of a bsAb by developing an UHPLC SEC-UV method combined with native ESI-MS for the simultaneous quantification and identification of size variants (Haberger, Leiss, et al., 2016). As shown for stability and bio-process samples, LC-nMS did not only facilitate confident peak assignment by determining the accurate mass but also was capable of detecting unknown, low-abundant species, which were chromatographically indistinguishable. High sensitivity for the identification of mispaired variants is of particular relevance for the analysis of highly potent bio-therapeutics, such as bispecific T-cell engager antibodies or Cytokine-fusion proteins, as they can pose potential safety risks (Lee et al., 2019). For this purpose, Yan, Xing, Wang, Daly, and Li (2019) developed a method using online nMS detection for the quantification of homodimers. Leveraging the unique properties of the employed mixed-mode SEC column and therefore achieving chromatographic baseline separation, mispaired bsAb variants could be quantified at levels as low as 0.1% surpassing the sensitivity of conventional UV-based methods. Size-variant analysis of different antibody formats, including mAbs, bsAbs,

ADCs, and Fc-fusion proteins, by applying native SE-HPLC-MS (optionally also with ion mobility IM-MS) was also reported in further studies, substantiating the broad applicability of this approach (Duivelshof, Fekete, Guillarme, & D'Atri, 2019; Ehkirch, Hernandez-Alba, et al., 2018; Jiang et al., 2020; Vandenheede, Sandra, & Sandra, 2019). In another recent work (Phung et al., 2019; Phung et al., 2020), separation of light-chain mispaired/scrambled bsAb species was realized by the development of a weak cation exchange chromatography method coupled to MS detection using volatile eluent systems. Most notably, this workflow allowed for the discrimination between the correctly paired bsAb (H1L1/H2L2) and the light chain–scrambled molecule (H1L2/H2L1), rendering it a promising approach for resolving (near-) isobaric species, which usually cannot be achieved by SEC.

In addition to assessing bio-therapeutics according to charge and size distribution, HIC and RPLC techniques exploiting molecular hydrophobicity properties have also emerged as complementary auxiliary tools. These analytic techniques have gained importance for the characterization of typical CQAs of ADCs such as average drug–antibody ratio (DAR), drug–load distribution, and the proportion of unconjugated mAb (Fleming, 2020). While both techniques provide adequate resolution, through preservation of the native structure, HIC facilitates a sensitive assessment of conformational variants, which can otherwise be impaired by the organic solvent conditions commonly applied in RPLC. As HIC separation usually involves high initial concentrations of low volatile salts, detection is typically restricted to UV- and FLD-based approaches. More recent studies have, however, presented novel hydrophobic resins which allow the use of ammonium acetate at MS-compatible concentrations in combination with organic modifier(s) in the eluting mobile phase for the characterization of mAbs (Chen et al., 2018) and ADCs (Chen, Yang, et al., 2019). In both cases, good chromatographic resolution was achieved without sacrificing the conformational integrity as demonstrated by the native-like charge envelope distribution of the obtained MS spectra. An even more appealing approach was developed by Wei, Han, Tang, Sandoval, and Zhang (2019) for the separation of oxidation and free thiol variants of a mAb, utilizing isocratic chromatographic conditions at low concentrations of volatile salts (<250 nM ammonium acetate) without the need for organic modifiers, using a polypropyl A column. Finally, Yan, Xing, Wang, Daly, and Li (2020) proposed the incorporation of a postcolumn "makeup and splitting flow" platform, which allows for the modulation of the effluent by diluting the salt concentration and reducing the flow rate before entering the MS system. As demonstrated for the analysis of two mAb samples and one ADC mimic, this strategy displayed good performance in terms of separation efficiency and MS data quality while adhering to analytical-scale columns with well-established ligands.

In spite of the outlined capabilities offered by state-of-the-art LC-nMS workflows for different separation modes, broader implementation of this technology, especially in QC settings, still faces a number of challenges. From a technical perspective, major concerns include run-to-run and instrument-to-instrument variations of the mass spectrometers, data integrity and other instrument qualification aspects, method transferability between different instrument types, and the lack of dedicated software packages supporting fully automated data analysis. In addition, separation efficiency can be impaired by switching from standard eluent systems (nonvolatile and high ionic strength) to MS-compatible eluents, resulting in poor peak resolution and difficulties during method bridging, which was recently demonstrated in a systematic study of 24 mAbs, 2 ADCs, and 4 Fc-fusion proteins utilizing nonvolatile versus volatile buffers for SEC (Goyon, D'Atri, et al., 2017).

7.5 Two-/multidimensional LC as a complementary tool to characterize bio-therapeutics

Despite its versatility and significant improvements in robustness, implementation and application of LC in routine use for bio-therapeutics present several challenges. The effective separation of multicomponent mixtures by any liquid chromatographic approach is complicated by inherent technical limitations, such as restricted peak capacity (at reasonable run times), resolving power, and selectivity. To overcome these problems, harnessing two or more dimensions of chromatography (2D-LC, mD-LC) has emerged as a logical technological evolution. By transferring the effluent of the first dimension onto a second-dimension column, and thereby leveraging (at least partly) orthogonal separation modes, the ability to separate complex mixtures or modifications that coelute in the first dimension is greatly enhanced (Li, Stoll, & Carr, 2009). Traditionally, when orthogonal chromatography was required, the combination of various phases was achieved mostly separately, with time-consuming tedious fractionation steps in between (i.e., offline 2D-LC). Developments in online 2D-LC over recent years have demonstrated the numerous benefits especially in terms of sample throughput and flexibility (Pirok, Stoll, & Schoenmakers, 2019; Sandra, Steenbeke, Vandenheede, Vanhoenacker, & Sandra, 2017; Stoll & Carr, 2017). Due to technical advances and also the subsequent commercialization of purpose-designed instrumentation, 2D-LC applications have also found increasing use within the biopharmaceutical industry. A variety of studies have reported combinations of different separation modes spanning from affinity chromatography to SEC, IEC, HIC, and RPLC (Sandra et al., 2016; Williams, Read, Agarabi, Lute, & Brorson, 2017), some of which will be discussed later.

Based on the operation procedure, 2D-LC approaches can be divided into two major categories, namely, comprehensive 2D-LC (LC × LC) and (multiple) heart cutting 2D-LC [(m)LC−LC]. The first, where each eluting fraction from the first dimension is directly transferred into the second, is predominantly used for omics-type applications (e.g., metabolomics, host cell protein analysis) (Doneanu et al., 2012; Willmann et al., 2015) and extensive comparative exercises as, for example, required for the characterization of biosimilar candidates relative to the respective originator molecule (Sorensen et al., 2016). On the other hand, (m)LC−LC is conceptualized to enable isolation of selected target compounds and subsequent in-depth characterization of specific peaks occurring during routine analysis. After selecting the fraction(s) of interest in the first dimension, the analyte(s) can be processed in various ways.

Alternative to offering an orthogonal separation mode, the second dimension can also serve as an effective online desalting/buffer exchange tool to mediate between nonvolatile eluents usually employed in the first dimension (e.g., SEC, IEC, and HIC) and MS detection (Luo et al., 2017; Petersson, Haselmann, & Buckenmaier, 2016; Stoll et al., 2015). To achieve this, the second dimension is usually operated using reversed-phase stationary phases combined with denaturing organic solvent systems (Birdsall et al., 2015; Largy, Catrain, Van Vyncht, & Delobel, 2016; Sarrut et al., 2016) or, more recently, by employing SEC chromatography combined with nondenaturing aqueous buffer solvent systems, whereby preserving the native structure. Direct hyphenation of native separations to MS was historically often hampered by poor chromatographic separation, as a result of the weak ionic and pH strengths of MS-compatible mobile phases (Goyon, D'Atri, et al., 2017). In this case, 2D-LC offers a potential workaround by utilizing well-established yet often MS-incompatible mobile phases in the first dimension followed by a fast desalting step to allow direct MS-coupling without sacrificing chromatographic performance. According to this idea, Ehkirch et al. developed an online SEC × SEC methodology (hyphenated to IM-MS) (Birrer et al., 2019) that demonstrated the benefits of applying routine SEC conditions in the first dimension for a variety of mAbs featuring acidic, basic, or hydrophobic properties. The same authors also reported on a similar approach for the characterization of ADCs using HIC × SEC (Ehkirch, D'Atri, et al., 2018).

Highly sophisticated and customized multidimensional LC (mD-LC) instrument setups can facilitate further sample manipulation, such as online reduction and digestion and, thus, allow extensive characterization of a single peak by combining bottom-up, subunit, and intact mass analysis. In this context, Bathke, Klemm, Gstöttner, Bell, and Kopf (2018) introduced a fully automated LC system enabling online sampling, reduction, chromatographic separation, and successive analysis of reduced heavy and light chains by MS detection. Owing to the flexibility to use different separation modes and run project-specific methods in the first dimension, the authors stated that this system is ideally suited for unveiling unknown peaks observed during QC measurements. Gstottner et al. (2018) proposed a further extension of this setup by integrating an immobilized trypsin cartridge for online protein digestion resulting in a four-dimensional (4D) LC–MS method. In this work, a proof-of-concept study, conducted for charge variant analysis of a therapeutic antibody demonstrated good comparability with the traditional offline fractionation process with respect to the deamidation, oxidation, and glycation variants investigated. More recently, a similar approach was adapted for the characterization of ADCs switching between a 2D-LC (SEC–HIC) and 4D-LC–MS workflow (SEC–reduction–digestion–RPLC) (Goyon et al., 2019). While the 2D-LC method was restricted to the analysis of the monomer fraction, online sample denaturation and reduction as applied in the 4D-LC–MS setup enabled determination of the average DAR for all peaks including dimers and higher order aggregates. Combined with the subsequent peptide mapping analysis, this multilayer approach provides simultaneous information about the DAR as well as the location/amount of PTMs for a given size-variant species. The versatility of state-of-the-art mD-LC systems was further highlighted by Camperi and coworkers in recent studies, who demonstrated the implementation of an online 2D (IdeS digestion–RPLC–MS) workflow for subunit analysis (Camperi, Guillarme, Lei, & Stella, 2020), an online 3D (IdeS digestion–reduction RPLC–HILIC–MS) workflow for evaluating glycosylation profiles (Camperi, Dai, Guillarme, & Stella, 2020a) and combination of an online 3D (protein-A–reduction–RPLC–HILIC–MS) and 4D (protein-A–reduction

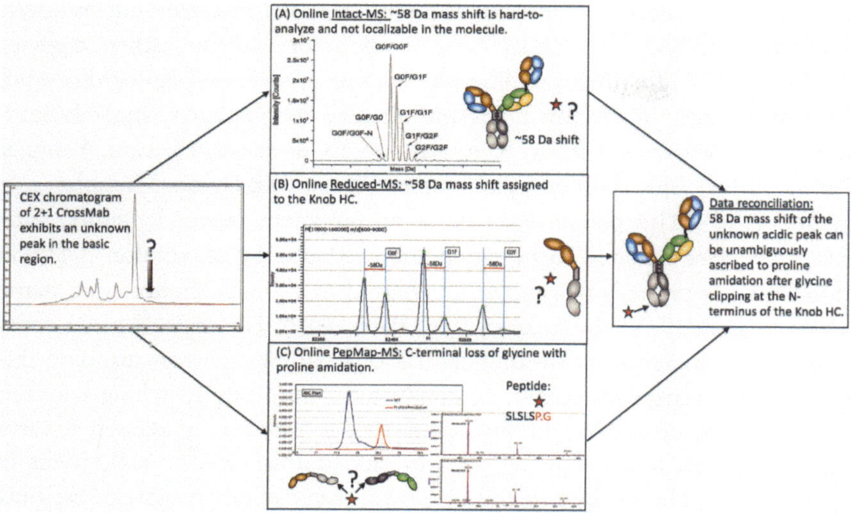

Figure 7.2 Characterization of product variants at varying organizational levels by mD-LC–MS. The investigated peak (indicated by the *red arrow*) of a bispecific 2 + 1 CrossMab format exhibits modifications (indicated as *red star*) at the N-terminus of the Knob HC. In this case, LC–MS-based peptide mapping would be unable to determine the space-resolved position of this modification, since both HCs incorporate the same peptide (SLSLSPG). Combining these results with the analysis at the subunit level after disulfide reduction affords discrimination between both proteoforms and ultimately leads to unambiguous assignment of the PTMs. In contrast to offline workflows, peak characterization by mD-LC–MS can be confidently achieved without the need for sample fractionation prior to analysis. *LC*, Liquid chromatography; *MS*, mass spectrometry.

RPLC–on-column tryptic digestion–RPLC–MS/MS) workflow for assessing mAb variants from harvested cell culture fluid (Camperi, Dai, Guillarme, & Stella, 2020b).

The growing number of published studies involving in-depth characterization of bio-therapeutics by mD-LC–MS highlights the advantages of the approach compared to traditional offline peak assignment (Bathke et al., 2018; Gstottner et al., 2018). As conventional offline approaches comprise a multitude of manual and time-consuming steps prior to MS analysis, such as peak fractionation, buffer exchange, re-chromatography, and in-solution tryptic digestion, they are almost exclusively performed for bio-therapeutics in later stages of development (Gandhi et al., 2012; Miao et al., 2017). In contrast, implementation of mD-LC–MS methods holds promise for fast, robust, and fully automated sample processing requiring minimal material and resources, and may thus become a conceivable solution for even in earlier stages of development. Furthermore, traditionally applied in-solution peptide digestion, especially at long incubation times and elevated temperatures, arguably leads to chemical degradation (Hao, Ren, Datta, Tam, & Sze, 2015; Lam, Yang, & Cleland, 1997) and adsorption phenomena (Warwood, Byron, Humphries, & Knight, 2013). This potential source of measurement bias and method variability has been reported to be significantly reduced by on-column digestion approaches, as proteolysis is typically achieved in a much shorter time frame (Toth et al., 2017; Toth, Kuklenyik, & Barr, 2019). However, a more comprehensive comparison of in-solution and on-column digestion spanning distinct protein formats remains to be conducted. Finally, by switching between different operation modes, the peak of interest can be readily assessed at varying organization levels allowing peptide- and domain-specific assignment of modifications. The benefits of this proposed characterization strategy are outlined in Fig. 7.2 for the analysis of a 2 + 1 CrossMAb format. Boosted by advances in modulating solvents to address compatibility issues between different dimensions (Chen, Li, & Schmitz, 2019; Stoll, Shoykhet, Petersson, & Buckenmaier, 2017) and the advent of new enzymatic columns, it is easily conceivable that novel application fields for the characterization of biotherapeutics will be identified in the near future. Apart from the analytical capabilities provided by 2D/mD-LC−MS approaches, the practicability of deploying elaborate, often specifically configured complex instrumentation setups has to be evaluated in terms of robustness, ease of use, and method transferability. Commercialized instrument platforms and appropriate software solutions offering the required flexibility to perform this kind of experiments—but also meeting the demands from a quality laboratory perspective—would certainly foster a broader implementation of this technique within the biopharmaceutical industry.

7.6 Conclusion

Conventional QC- and extended characterization analytical methods applied in the biopharmaceutical industry, established with the respective manufacturing expectations concerning method performance (e.g., ease of use, robustness), face increasing difficulties to cover all relevant product-related quality attributes of new bio-therapeutic formats. These limitations primarily stem from the relatively low resolution of chromatographic and CE-based approaches coupled to UV/FLD, which impede characterization

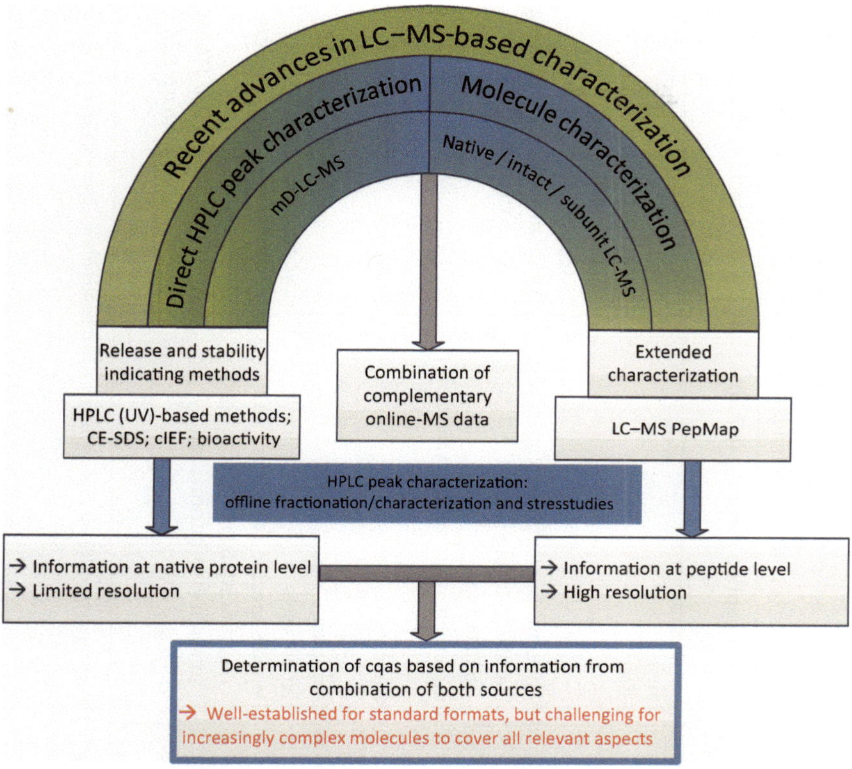

Figure 7.3 Complementary strategy for the characterization of complex formats. State-of-the-art LC–MS techniques, including native LC–MS and mD-LC–MS, are poised to fill the knowledge gap remained unaddressed after conventional quality testing by providing an additional layer of information. *LC*, Liquid chromatography; *MS*, mass spectrometry.

of products featuring an inherently higher level of complexity and heterogeneity, arising from additional product size and charge variants. Peptide-based MAM as an emerging technology has the potential to remedy some of these problems by offering enhanced sensitivity and selectivity for the simultaneous detection and quantification of multiple product attributes. In contrast to the analysis of the intact molecule, however, protein digestion leads to the loss of higher order structure information preventing the description of the proteoform distribution, which becomes of particular relevance for novel bio-therapeutic exceeding the complexity of standard (IgG-like) formats. In this regard, a combination of LC-nMS and 2D/mD-LC–MS-based approaches capable of subunit and intact mass measurement

Table 7.3 Comparison between the herein described approaches for assessing product attributes.

Category	Separation-based methods coupled to UV/FLD	LC–MS peptide mapping (MAM)	Online intact LC–MS approaches
Resolution	Low	High	Medium
Sensitivity	Low	High	Medium (depending on chromatographic separation)
Robustness	High (simple sample preparation and highly robust instrumentations)	Moderate (susceptible to artificial artifacts during sample processing)	Medium to high (depending on instrument robustness)
GMP compliance	Well established	Implementation under evaluation	Not established
Sample throughput	High (after method is characterized)	Medium	Medium
Data evaluation	Simple	Complex	Complex
Operator	Trained personnel	Expert users	Expert users
Derived information	Analysis under native conditions, but no direct identification of product attributes (Characterization method needed to identify peaks)	Peptide-specific assignment and quantification of PTMs Loss of intact/native information	Proteoform analysis at native conditions Peptide-specific location of PTMs not attainable (native MS)

FLD, Fluorescence detection; *GMP*, good manufacturing practice; *LC*, liquid chromatography; *MS*, mass spectrometry; *PTM*, posttranslational modification; *UV*, ultraviolet.

complement the well-established techniques and, therefore, facilitate the comprehensive characterization of these molecules. As discussed here in detail, direct coupling of different LC modes to MS while maintaining the higher order structure or, at least, conserving its information, achieves confident identification and quantification of proteoforms that may remain undetected by conventional UV-based methods or LC–MS peptide mapping. More specifically, native SE-HPLC-MS is capable of resolving and identifying a large variety of related size variants (as exemplarily illustrated in Fig. 7.1B for a bispecific 1 + 1 CrossMAb format), even if the single species are—unlike for standard IgG-like formats—chromatographically inseparable. In addition, analysts can benefit from combining results obtained by MAM, native IE-HPLC-MS, and 2D/mD-LC–MS approaches to unveil complex, overlapping peak patterns, which can be frequently observed during charge variant analysis of complex molecules. The proposed role of these structure-preserving approaches for the in-depth characterization of innovative bio-therapeutic formats is illustrated in Fig. 7.3. Similar to MAM, implementation of MS-based techniques, relying on complex instrumentations, specifically trained experts and dedicated software packages, into a QC environment needs to be carefully evaluated and established with respect to current regulatory guidelines. The major benefits and challenges/drawbacks of all required strategies forming the basis for future product quality attribute testing are summarized in Table 7.3.

In summary, native LC–MS and mD-LC–MS techniques show potential to complement the existing analytical toolbox for quality testing and are expected to play a crucial role for control strategies of next-generation bio-therapeutics.

References

Alt, N., Zhang, T. Y., Motchnik, P., Taticek, R., Quarmby, V., Schlothauer, T., ... Harris, R. J. (2016). Determination of critical quality attributes for monoclonal antibodies using quality by design principles. *Biologicals: Journal of the International Association of Biological Standardization*, *44*(5), 291–305.

Auclair, J., Rathore, A., & Krull, I. S. (2019). Methods and purposes for determining higher order structures of biopharmaceuticals. *LCGC North America*, *37*(1), 34–42.

Bailey, A. O., Han, G., Phung, W., Gazis, P., Sutton, J., Josephs, J. L., & Sandoval, W. (2018). Charge variant native mass spectrometry benefits mass precision and dynamic range of monoclonal antibody intact mass analysis. *MAbs*, *10*(8), 1214–1225.

Bathke, A., Klemm, D., Gstöttner, C., Bell, C., & Kopf, R. (2018). Rapid online reduction and characterization of protein modifications using fully automated two-dimensional high performance liquid chromatography–mass spectrometry. *LCGC Europe*, *31*(10–21).

Beck, A., & Liu, H. (2019). Macro- and micro-heterogeneity of natural and recombinant IgG antibodies. *Antibodies (Basel)*, *8*(1).

Beck, A., Terral, G., Debaene, F., Wagner-Rousset, E., Marcoux, J., Janin-Bussat, M. C., ... Cianferani, S. (2016). Cutting-edge mass spectrometry methods for the multi-level structural characterization of antibody-drug conjugates. *Expert Review of Proteomics*, *13* (2), 157–183.

Beck, A., Wagner-Rousset, E., Ayoub, D., Van Dorsselaer, A., & Sanglier-Cianferani, S. (2013). Characterization of therapeutic antibodies and related products. *Analytical Chemistry*, *85*(2), 715–736.

Belov, A. M., Viner, R., Santos, M. R., Horn, D. M., Bern, M., Karger, B. L., & Ivanov, A. R. (2017). Analysis of proteins, protein complexes, and organellar proteomes using sheathless capillary zone electrophoresis—Native mass spectrometry. *Journal of the American Society for Mass Spectrometry*, *28*(12), 2614–2634.

Bereszczak, J. Z., Barbu, I. M., Tan, M., Xia, M., Jiang, X., van Duijn, E., & Heck, A. J. (2012). Structure, stability and dynamics of norovirus P domain derived protein complexes studied by native mass spectrometry. *Journal of Structural Biology*, *177*(2), 273–282.

Berger, S., & Fredette, J. (2017). Mass spectrometry in late development and QC: Practical considerations for multi-attribute monitoring and beyond. *LCGC - The column*, *13*(9), 9–13.

Bertolotti-Ciarlet, A., Wang, W., Lownes, R., Pristatsky, P., Fang, Y., McKelvey, T., ... Vlasak, J. (2009). Impact of methionine oxidation on the binding of human IgG1 to Fc Rn and Fc gamma receptors. *Molecular Immunology*, *46*(8–9), 1878–1882.

Birdsall, R. E., Shion, H., Kotch, F. W., Xu, A., Porter, T. J., & Chen, W. (2015). A rapid on-line method for mass spectrometric confirmation of a cysteine-conjugated antibody-drug-conjugate structure using multidimensional chromatography. *MAbs*, *7* (6), 1036–1044.

Birrer, M. J., Moore, K. N., Betella, I., & Bates, R. C. (2019). Antibody-drug conjugate-based therapeutics: State of the science. *Journal of the National Cancer Institute*, *111*.

Blake-Haskins, A. W., Lin, Y.-H., Wu, Z., Perkins, M. D., & Spitznagel, T. M. (2014). *Biophysical and structural characterization needed prior to proof of concept. Biophysical methods for biotherapeutics* (pp. 93–123). John Wiley & Sons.

Bonnington, L., Lindner, I., Gilles, U., Kailich, T., Reusch, D., & Bulau, P. (2017). Application of hydrogen/deuterium exchange-mass spectrometry to biopharmaceutical development requirements: Improved sensitivity to detection of conformational changes. *Analytical Chemistry*, *89*(16), 8233–8237.

Borisov, O. V., Alvarez, M., Carroll, J. A., & Brown, P. W. (2015). Sequence variants and sequence variant analysis in biotherapeutic proteins. *ACS Symposium Series*, *1201*(2), 63–117.

Boyd, D., Kaschak, T., & Yan, B. (2011). HIC resolution of an IgG1 with an oxidized Trp in a complementarity determining region. *Journal of Chromatography. B, Analytical Technologies in the Biomedical and Life Sciences*, *879*(13–14), 955–960.

Brinson, R. G., Arbogast, L. W., Marino, J. P., & Delaglio, F. (2020). Best practices in utilization of 2D-NMR spectral data as the input for chemometric analysis in biopharmaceutical applications. *Journal of Chemical Information and Modeling*, *60*(4), 2339–2355.

Brodbelt, J. S. (2016). Ion activation methods for peptides and proteins. *Analytical Chemistry*, *88*(1), 30–51.

Buettner, A., Maier, M., Bonnington, L., Bulau, P., & Reusch, D. (2020). Multi-attribute monitoring of complex erythropoetin beta glycosylation by GluC liquid chromatography-mass spectrometry peptide mapping. *Analytical Chemistry*, *92*(11), 7574–7580.

Byrne, H., Conroy, P. J., Whisstock, J. C., & O'Kennedy, R. J. (2013). A tale of two specificities: Bispecific antibodies for therapeutic and diagnostic applications. *Trends in Biotechnology*, *31*(11), 621–632.

Camperi, J., Dai, L., Guillarme, D., & Stella, C. (2020a). Development of a 3D-LC/MS workflow for fast, automated, and effective characterization of glycosylation patterns of biotherapeutic products. *Analytical Chemistry*, *92*(6), 4357–4363.

Camperi, J., Dai, L., Guillarme, D., & Stella, C. (2020b). Fast and automated characterization of monoclonal antibody minor variants from cell cultures by combined protein-A and multidimensional LC/MS methodologies. *Analytical Chemistry*, *92*(12), 8506–8513.

Camperi, J., Guillarme, D., Lei, M., & Stella, C. (2020). Automated middle-up approach for the characterization of biotherapeutic products by combining on-line hinge-specific digestion with RPLC-HRMS analysis. *Journal of Pharmaceutical and Biomedical Analysis*, *182*, 113130.

Cao, M., De Mel, N., Shannon, A., Prophet, M., Wang, C., Xu, W., ... Wang, J. (2019). Charge variants characterization and release assay development for co-formulated antibodies as a combination therapy. *MAbs*, *11*(3), 489–499.

Chames, P., Van Regenmortel, M., Weiss, E., & Baty, D. (2009). Therapeutic antibodies: Successes, limitations and hopes for the future. *British Journal of Pharmacology*, *157*(2), 220–233.

Chen, B., Lin, Z., Alpert, A. J., Fu, C., Zhang, Q., Pritts, W. A., & Ge, Y. (2018). Online hydrophobic interaction chromatography-mass spectrometry for the analysis of intact monoclonal antibodies. *Analytical Chemistry*, *90*(12), 7135–7138.

Chen, B., Peng, Y., Valeja, S. G., Xiu, L., Alpert, A. J., & Ge, Y. (2016). Online hydrophobic interaction chromatography-mass spectrometry for top-down proteomics. *Analytical Chemistry*, *88*(3), 1885–1891.

Chen, T. H., Yang, Y., Zhang, Z., Fu, C., Zhang, Q., Williams, J. D., & Wirth, M. J. (2019). Native reversed-phase liquid chromatography: A technique for LCMS of intact antibody-drug conjugates. *Analytical Chemistry*, *91*(4), 2805–2812.

Chen, Y., Li, J., & Schmitz, O. J. (2019). Development of an at-column dilution modulator for flexible and precise control of dilution factors to overcome mobile phase incompatibility in comprehensive two-dimensional liquid chromatography. *Analytical Chemistry*, *91*.

Chung, C. H., Mirakhur, B., Chan, E., Le, Q. T., Berlin, J., Morse, M., ... Platts-Mills, T. A. (2008). Cetuximab-induced anaphylaxis and IgE specific for galactose-alpha-1,3-galactose. *The New England Journal of Medicine*, *358*(11), 1109–1117.

Cole, S. R., Ma, X., Zhang, X., & Xia, Y. (2012). Electron transfer dissociation (ETD) of peptides containing intrachain disulfide bonds. *Journal of the American Society for Mass Spectrometry*, *23*(2), 310–320.

Darula, Z., Pap, A., & Medzihradszky, K. F. (2019). Extended sialylated *O*-glycan repertoire of human urinary glycoproteins discovered and characterized using electron-transfer/higher-energy collision dissociation. *Journal of Proteome Research*, *18*(1), 280–291.

de Haan, N., Falck, D., & Wuhrer, M. (2019). Monitoring of immunoglobulin *N*- and *O*-glycosylation in health and disease. *Glycobiology*, *30*.

del Val, I. J., Kontoravdi, C., & Nagy, J. M. (2010). Towards the implementation of quality by design to the production of therapeutic monoclonal antibodies with desired glycosylation patterns. *Biotechnology Progress*, *26*(6), 1505–1527.

Dick, L. W., Jr., Mahon, D., Qiu, D., & Cheng, K. C. (2009). Peptide mapping of therapeutic monoclonal antibodies: Improvements for increased speed and fewer artifacts. *Journal of Chromatography B: Analytical Technologies in the Biomedical and Life Sciences*, *877*(3), 230–236.

Diepold, K., Bomans, K., Wiedmann, M., Zimmermann, B., Petzold, A., Schlothauer, T., ... Bulau, P. (2012). Simultaneous assessment of Asp isomerization and Asn deamidation in recombinant antibodies by LC-MS following incubation at elevated temperatures. *PLoS One*, 7(1), e30295.

Doneanu, C. E., Xenopoulos, A., Fadgen, K., Murphy, J., Skilton, S. J., Prentice, H., ... Chen, W. (2012). Analysis of host-cell proteins in biotherapeutic proteins by comprehensive online two-dimensional liquid chromatography/mass spectrometry. *MAbs*, *4* (1), 24–44.

Donnelly, D. P., Rawlins, C. M., DeHart, C. J., Fornelli, L., Schachner, L. F., Lin, Z., ... Agar, J. N. (2019). Best practices and benchmarks for intact protein analysis for top-down mass spectrometry. *Nature Methods*, *16*(7), 587–594.

Duivelshof, B. L., Fekete, S., Guillarme, D., & D'Atri, V. (2019). A generic workflow for the characterization of therapeutic monoclonal antibodies-application to daratumumab. *Analytical and Bioanalytical Chemistry*, *411*(19), 4615–4627.

Ehkirch, A., D'Atri, V., Rouviere, F., Hernandez-Alba, O., Goyon, A., Colas, O., ... Cianferani, S. (2018). An online four-dimensional HICxSEC-IMxMS methodology for proof-of-concept characterization of antibody drug conjugates. *Analytical Chemistry*, *90*(3), 1578–1586.

Ehkirch, A., Goyon, A., Hernandez-Alba, O., Rouviere, F., D'Atri, V., Dreyfus, C., ... Cianferani, S. (2018). A novel online four-dimensional SECxSEC-IMxMS methodology for characterization of monoclonal antibody size variants. *Analytical Chemistry*, *90* (23), 13929–13937.

Ehkirch, A., Hernandez-Alba, O., Colas, O., Beck, A., Guillarme, D., & Cianferani, S. (2018). Hyphenation of size exclusion chromatography to native ion mobility mass spectrometry for the analytical characterization of therapeutic antibodies and related products. *Journal of Chromatography. B, Analytical Technologies in the Biomedical and Life Sciences*, *1086*, 176–183.

EMA (2020). *European Medicines Agency—Medicines, 2020.* Available at https://www.ema.europa.eu/en/medicines (Accessed: 25 September 2020).

Eon-Duval, A., Broly, H., & Gleixner, R. (2012). Quality attributes of recombinant therapeutic proteins: An assessment of impact on safety and efficacy as part of a quality by design development approach. *Biotechnology Progress*, *28*(3), 608–622.

Evens, R., & Kaitin, K. (2015). The evolution of biotechnology and its impact on health care. *Health Affairs (Millwood)*, *34*(2), 210–219.

Evens, R. P. (2016). Pharma success in product development—Does biotechnology change the paradigm in product development and attrition. *The AAPS Journal*, *18*(1), 281–285.

Faid, V., Leblanc, Y., Bihoreau, N., & Chevreux, G. (2018). Middle-up analysis of monoclonal antibodies after combined IgdE and IdeS hinge proteolysis: Investigation of free sulfhydryls. *Journal of Pharmaceutical and Biomedical Analysis*, *149*, 541–546.

Fan, G., Wang, Z., Hao, M., & Li, J. (2015). Bispecific antibodies and their applications. *Journal of Hematology & Oncology*, *8*, 130.

FDA (2020). *New drugs at FDA: CDER's new molecular entities and new therapeutic biological products, 2020.* Available at https://www.fda.gov/drugs/development-approval-process-drugs/new-drugs-fda-cders-new-molecular-entities-and-new-therapeutic-biological-products (Accessed: 25 September 2020).

Fleming, R. (2020). ADC analysis by hydrophobic interaction chromatography. *Methods in Molecular Biology*, *2078*, 147–161.

Folzer, E., Diepold, K., Bomans, K., Finkler, C., Schmidt, R., Bulau, P., ... Koulov, A. V. (2015). Selective oxidation of methionine and tryptophan residues in a therapeutic IgG1 molecule. *Journal of Pharmaceutical Sciences*, *104*(9), 2824–2831.

Fornelli, L., Damoc, E., Thomas, P. M., Kelleher, N. L., Aizikov, K., Denisov, E., ... Tsybin, Y. O. (2012). Analysis of intact monoclonal antibody IgG1 by electron transfer dissociation Orbitrap FTMS. *Molecular & Cellular Proteomics: MCP, 11*(12), 1758–1767.

Fussl, F., Cook, K., Scheffler, K., Farrell, A., Mittermayr, S., & Bones, J. (2018). Charge variant analysis of monoclonal antibodies using direct coupled pH gradient cation exchange chromatography to high-resolution native mass spectrometry. *Analytical Chemistry, 90*(7), 4669–4676.

Fussl, F., Trappe, A., Cook, K., Scheffler, K., Fitzgerald, O., & Bones, J. (2019). Comprehensive characterisation of the heterogeneity of adalimumab via charge variant analysis hyphenated on-line to native high resolution Orbitrap mass spectrometry. *MAbs, 11*(1), 116–128.

Gabrielson, J. P., & Weiss, W. F. T. (2015). Technical decision-making with higher order structure data: Starting a new dialogue. *Journal of Pharmaceutical Sciences, 104*(4), 1240–1245.

Gandhi, S., Ren, D., Xiao, G., Bondarenko, P., Sloey, C., Ricci, M. S., & Krishnan, S. (2012). Elucidation of degradants in acidic peak of cation exchange chromatography in an IgG1 monoclonal antibody formed on long-term storage in a liquid formulation. *Pharmaceutical Research, 29*(1), 209–224.

Goyon, A., Beck, A., Colas, O., Sandra, K., Guillarme, D., & Fekete, S. (2017). Evaluation of size exclusion chromatography columns packed with sub-3mum particles for the analysis of biopharmaceutical proteins. *Journal of Chromatography A, 1498*, 80–89.

Goyon, A., D'Atri, V., Colas, O., Fekete, S., Beck, A., & Guillarme, D. (2017). Characterization of 30 therapeutic antibodies and related products by size exclusion chromatography: Feasibility assessment for future mass spectrometry hyphenation. *Journal of Chromatography B: Analytical Technologies in the Biomedical and Life Sciences, 1065–1066*, 35–43.

Goyon, A., Kim, M., Dai, L., Cornell, C., Jacobson, F., Guillarme, D., & Stella, C. (2019). Streamlined characterization of an antibody-drug conjugate by two-dimensional and four-dimensional liquid chromatography/mass spectrometry. *Analytical Chemistry, 91*.

Grandori, R. (2003). Origin of the conformation dependence of protein charge-state distributions in electrospray ionization mass spectrometry. *Journal of Mass Spectrometry: JMS, 38*(1), 11–15.

Grassi, L., & Cabrele, C. (2019). Susceptibility of protein therapeutics to spontaneous chemical modifications by oxidation, cyclization, and elimination reactions. *Amino Acids, 51*(10–12), 1409–1431.

Gstottner, C., Klemm, D., Haberger, M., Bathke, A., Wegele, H., Bell, C., & Kopf, R. (2018). Fast and automated characterization of antibody variants with 4D HPLC/MS. *Analytical Chemistry, 90*(3), 2119–2125.

Gunasekaran, K., Pentony, M., Shen, M., Garrett, L., Forte, C., Woodward, A., ... Yan, W. (2010). Enhancing antibody Fc heterodimer formation through electrostatic steering effects: Applications to bispecific molecules and monovalent IgG. *The Journal of Biological Chemistry, 285*(25), 19637–19646.

Haberger, M., Bomans, K., Diepold, K., Hook, M., Gassner, J., Schlothauer, T., ... Bulau, P. (2014). Assessment of chemical modifications of sites in the CDRs of recombinant antibodies: Susceptibility vs. functionality of critical quality attributes. *MAbs, 6*(2), 327–339.

Haberger, M., Bonnington, L., Bomans, K., Maier, M., Eltner, S., Dembowski, M., ... Bulau, P. (2016). Application of electrospray ionization mass spectrometry in a quality control laboratory. *American Pharmaceutical Review.*

Haberger, M., Leiss, M., Heidenreich, A. K., Pester, O., Hafenmair, G., Hook, M., ... Bulau, P. (2016). Rapid characterization of biotherapeutic proteins by size-exclusion chromatography coupled to native mass spectrometry. *MAbs, 8*(2), 331–339.

Hagihara, Y., & Saerens, D. (2014). Engineering disulfide bonds within an antibody. *Biochimica et Biophysica Acta, 1844*(11), 2016–2023.

Hao, P., Ren, Y., Datta, A., Tam, J. P., & Sze, S. K. (2015). Evaluation of the effect of trypsin digestion buffers on artificial deamidation. *Journal of Proteome Research, 14*(2), 1308–1314.

Harris, R. J. (2005). *Heterogeneity of recombinant antibodies: Linking structure to function*, . *Developments in Biologicals (Basel)* (122, pp. 117–127).

Haverick, M., Mengisen, S., Shameem, M., & Ambrogelly, A. (2014). Separation of mAbs molecular variants by analytical hydrophobic interaction chromatography HPLC: Overview and applications. *MAbs, 6*(4), 852–858.

Higel, F., Seidl, A., Sorgel, F., & Friess, W. (2016). *N*-Glycosylation heterogeneity and the influence on structure, function and pharmacokinetics of monoclonal antibodies and Fc fusion proteins. *European Journal of Pharmaceutics and Biopharmaceutics: Official Journal of Arbeitsgemeinschaft fur Pharmazeutische Verfahrenstechnik e.V, 100*, 94–100.

Houde, D., & Engen, J. R. (2013). Conformational analysis of recombinant monoclonal antibodies with hydrogen/deuterium exchange mass spectrometry. *Methods in Molecular Biology, 988*, 269–289.

Huang, R. Y., & Chen, G. (2014). Higher order structure characterization of protein therapeutics by hydrogen/deuterium exchange mass spectrometry. *Analytical and Bioanalytical Chemistry, 406*(26), 6541–6558.

ICH. (1999). *International Council for Harmonisation of technical requirements for pharmaceuticals for human use—Specifications: Test procedures and acceptance criteria for biotechnological/biological products Q6B*.

Jefferis, R. (2016). Posttranslational modifications and the immunogenicity of biotherapeutics. *Journal of Immunology Research, 2016*, 5358272.

Jenkins, N., Murphy, L., & Tyther, R. (2008). Post-translational modifications of recombinant proteins: Significance for biopharmaceuticals. *Molecular Biotechnology, 39*(2), 113–118.

Jennewein, M. F., & Alter, G. (2017). The immunoregulatory roles of antibody glycosylation. *Trends in Immunology, 38*(5), 358–372.

Jiang, H., Xu, W., Liu, R., Gupta, B., Kilgore, B., Du, Z., & Yang, X. (2020). Characterization of bispecific antibody production in cell cultures by unique mixed mode size exclusion chromatography. *Analytical Chemistry, 92*(13), 9312–9321.

Karst, D. J., Steinebach, F., & Morbidelli, M. (2018). Continuous integrated manufacturing of therapeutic proteins. *Current Opinion in Biotechnology, 53*, 76–84.

Kesik-Brodacka, M. (2018). Progress in biopharmaceutical development. *Biotechnology and Applied Biochemistry, 65*(3), 306–322.

Kim, S., Song, J., Park, S., Ham, S., Paek, K., Kang, M., ... Flores, M. (2017). Drifts in ADCC-related quality attributes of Herceptin(R): Impact on development of a trastuzumab biosimilar. *MAbs, 9*(4), 704–714.

Kitagawa, T. (1992). Investigation of higher order structures of proteins by ultraviolet resonance Raman spectroscopy. *Progress in Biophysics and Molecular Biology, 58*(1), 1–18.

Klein, C., Schaefer, W., & Regula, J. T. (2016). The use of CrossMAb technology for the generation of bi- and multispecific antibodies. *MAbs, 8*(6), 1010–1020.

Konijnenberg, A., Butterer, A., & Sobott, F. (2013). Native ion mobility-mass spectrometry and related methods in structural biology. *Biochimica et Biophysica Acta, 1834*(6), 1239–1256.

Kuhne, F., Bonnington, L., Malik, S., Thomann, M., Avenal, C., Cymer, F., ... Bulau, P. (2019). The impact of immunoglobulin G1 Fc sialylation on backbone amide H/D exchange. *Antibodies (Basel)*, *8*(4).

Labrijn, A. F., Janmaat, M. L., Reichert, J. M., & Parren, P. (2019). Bispecific antibodies: A mechanistic review of the pipeline. *Nature Reviews. Drug Discovery*, *18*(8), 585–608.

Lakbub, J. C., Shipman, J. T., & Desaire, H. (2018). Recent mass spectrometry-based techniques and considerations for disulfide bond characterization in proteins. *Analytical and Bioanalytical Chemistry*, *410*(10), 2467–2484.

Lam, X. M., Yang, J. Y., & Cleland, J. L. (1997). Antioxidants for prevention of methionine oxidation in recombinant monoclonal antibody HER2. *Journal of Pharmaceutical Sciences*, *86*(11), 1250–1255.

Largy, E., Catrain, A., Van Vyncht, G., & Delobel, A. (2016). 2D-LC–MS for the analysis of monoclonal antibodies and antibody–drug conjugates in a regulated environment. *spectroscopyonline.com*, *14*(2), 29–35.

Leblanc, Y., Bihoreau, N., & Chevreux, G. (2018). Characterization of human serum albumin isoforms by ion exchange chromatography coupled on-line to native mass spectrometry. *Journal of Chromatography. B, Analytical Technologies in the Biomedical and Life Sciences*, *1095*, 87–93.

Leblanc, Y., Ramon, C., Bihoreau, N., & Chevreux, G. (2017). Charge variants characterization of a monoclonal antibody by ion exchange chromatography coupled online to native mass spectrometry: Case study after a long-term storage at +5 degrees C. *Journal of Chromatography. B, Analytical Technologies in the Biomedical and Life Sciences*, *1048*, 130–139.

Lee, H. Y., Contreras, E., Register, A. C., Wu, Q., Abadie, K., Garcia, K., ... Jiang, G. (2019). Development of a bioassay to detect T-cell-activating impurities for T-cell-dependent bispecific antibodies. *Scientific Reports*, *9*(1), 3900.

Leney, A. C., & Heck, A. J. (2017). Native mass spectrometry: What is in the name? *Journal of the American Society for Mass Spectrometry*, *28*(1), 5–13.

Li, C. H., Nguyen, X., Narhi, L., Chemmalil, L., Towers, E., Muzammil, S., ... Jiang, Y. (2011). Applications of circular dichroism (CD) for structural analysis of proteins: Qualification of near- and far-UV CD for protein higher order structural analysis. *Journal of Pharmaceutical Sciences*, *100*(11), 4642–4654.

Li, P., Haque, M. A., & Blum, J. S. (2002). Role of disulfide bonds in regulating antigen processing and epitope selection. *Journal of Immunology*, *169*(5), 2444–2450.

Li, X., Stoll, D. R., & Carr, P. W. (2009). Equation for peak capacity estimation in two-dimensional liquid chromatography. *Analytical Chemistry*, *81*(2), 845–850.

Lin, J. C., Glover, Z. K., & Sreedhara, A. (2015). Assessing the utility of circular dichroism and FTIR spectroscopy in monoclonal-antibody comparability studies. *Journal of Pharmaceutical Sciences*, *104*(12), 4459–4466.

Loew, C., Knoblich, C., Fichtl, J., Alt, N., Diepold, K., Bulau, P., ... Grauschopf, U. (2012). Analytical protein a chromatography as a quantitative tool for the screening of methionine oxidation in monoclonal antibodies. *Journal of Pharmaceutical Sciences*, *101*(11), 4248–4257.

Lowe, D., Dudgeon, K., Rouet, R., Schofield, P., Jermutus, L., & Christ, D. (2011). Aggregation, stability, and formulation of human antibody therapeutics. *Advances in Protein Chemistry and Structural Biology*, *84*, 41–61.

Luo, H., Zhong, W., Yang, J., Zhuang, P., Meng, F., Caldwell, J., ... Welch, C. J. (2017). 2D-LC as an on-line desalting tool allowing peptide identification directly from MS unfriendly HPLC methods. *Journal of Pharmaceutical and Biomedical Analysis*, *137*, 139–145.

Ma, F., Raoufi, F., Bailly, M. A., Fayadat-Dilman, L., & Tomazela, D. (2020). Hyphenation of strong cation exchange chromatography to native mass spectrometry for high throughput online characterization of charge heterogeneity of therapeutic monoclonal antibodies. *MAbs*, *12*(1), 1763762.

Maas, C., Hermeling, S., Bouma, B., Jiskoot, W., & Gebbink, M. F. (2007). A role for protein misfolding in immunogenicity of biopharmaceuticals. *The Journal of Biological Chemistry*, *282*(4), 2229–2236.

Macchi, F. D., Yang, F., Li, C., Wang, C., Dang, A. N., Marhoul, J. C., ... Michels, D. A. (2015). Absolute quantitation of intact recombinant antibody product variants using mass spectrometry. *Analytical Chemistry*, *87*(20), 10475–10482.

Majumdar, R., Esfandiary, R., Bishop, S. M., Samra, H. S., Middaugh, C. R., Volkin, D. B., & Weis, D. D. (2015). Correlations between changes in conformational dynamics and physical stability in a mutant IgG1 mAb engineered for extended serum half-life. *MAbs*, 7(1), 84–95.

Marino, J. P., Brinson, R. G., Ladner, J. E., Gallagher, D. T., Arbogast, L. W., Huang, R. Y., ... Gallagher, E. S. (2015). Emerging technologies to assess the higher order structure of monoclonal antibodies. In J. E. Schiel, D. L. Davis, & O. V. Borisov (Eds.), *State-of-the-Art and Emerging Technologies for Therapeutic Monoclonal Antibody Characterization*. Washington, DC: American Chemical Society.

Miao, S., Xie, P., Zou, M., Fan, L., Liu, X., Zhou, Y., ... Tan, W. S. (2017). Identification of multiple sources of the acidic charge variants in an IgG1 monoclonal antibody. *Applied Microbiology and Biotechnology*, *101*(14), 5627–5638.

Misra, G. (2017). *Introduction to biomolecular structure and biophysics*. New York, Berlin Heidelberg: Springer.

Mouchahoir, T., & Schiel, J. E. (2018). Development of an LC-MS/MS peptide mapping protocol for the NISTmAb. *Analytical and Bioanalytical Chemistry*, *410*(8), 2111–2126.

Moussa, E. M., Panchal, J. P., Moorthy, B. S., Blum, J. S., Joubert, M. K., Narhi, L. O., & Topp, E. M. (2016). Immunogenicity of therapeutic protein aggregates. *Journal of Pharmaceutical Sciences*, *105*(2), 417–430.

Muneeruddin, K., Bobst, C. E., Frenkel, R., Houde, D., Turyan, I., Sosic, Z., & Kaltashov, I. A. (2017). Characterization of a PEGylated protein therapeutic by ion exchange chromatography with on-line detection by native ESI MS and MS/MS. *Analyst*, *142*(2), 336–344.

Nagarkar, R. P., Murphy, B. M., Yu, X., Manning, M. C., & Al-Azzam, W. A. (2013). Characterization of protein higher order structure using vibrational circular dichroism spectroscopy. *Current Pharmaceutical Biotechnology*, *14*(2), 199–208.

NIH (2020). *US National Library of Medicine—Clinical trials, 2020*. Available at https://clinicaltrials.gov/ (Accessed: 25 September 2020).

Pan, H., Chen, K., Chu, L., Kinderman, F., Apostol, I., & Huang, G. (2009). Methionine oxidation in human IgG2 Fc decreases binding affinities to protein A and FcRn. *Protein Science: A Publication of the Protein Society*, *18*(2), 424–433.

Petersson, P., Haselmann, K., & Buckenmaier, S. (2016). Multiple heart-cutting two dimensional liquid chromatography mass spectrometry: Towards real time determination of related impurities of bio-pharmaceuticals in salt based separation methods. *Journal of Chromatography A*, *1468*, 95–101.

Phung, W., Han, G., Polderdijk, S., Dillon, M., Shatz, W., Liu, P., ... Sandoval, W. (2019). Characterization of bispecific and mispaired IgGs by native charge-variant mass spectrometry. *International Journal of Mass Spectrometry*, *446*(116229).

Phung, W., Han, G., Polderdijk, S. G. I., Dillon, M., Shatz, W., Liu, P., ... Sandoval, W. (2020). Data on charge separation of bispecific and mispaired IgGs using native charge-variant mass spectrometry. *Data Brief*, *30*, 105435.

Pirok, B. W. J., Stoll, D. R., & Schoenmakers, P. J. (2019). Recent developments in two-dimensional liquid chromatography: Fundamental improvements for practical applications. *Analytical Chemistry*, *91*(1), 240–263.

Reusch, D., Haberger, M., Falck, D., Peter, B., Maier, B., Gassner, J., ... Wuhrer, M. (2015). Comparison of methods for the analysis of therapeutic immunoglobulin G Fc-glycosylation profiles—Part 2: Mass spectrometric methods. *MAbs*, 7(4), 732–742.

Reusch, D., Haberger, M., Kailich, T., Heidenreich, A. K., Kampe, M., Bulau, P., & Wuhrer, M. (2014). High-throughput glycosylation analysis of therapeutic immunoglobulin G by capillary gel electrophoresis using a DNA analyzer. *MAbs*, *6*(1), 185–196.

Reusch, D., Haberger, M., Maier, B., Maier, M., Kloseck, R., Zimmermann, B., ... Wuhrer, M. (2015). Comparison of methods for the analysis of therapeutic immunoglobulin G Fc-glycosylation profiles—Part 1: Separation-based methods. *MAbs*, 7(1), 167–179.

Roberts, C. J. (2014). Protein aggregation and its impact on product quality. *Current Opinion in Biotechnology*, *30*, 211–217.

Robinson, C. J., & Jones, C. (2011). Quality control and analytical techniques for biopharmaceuticals. *Future Science*, *3*(1), 81–95.

Rogers, R. S., Abernathy, M., Richardson, D. D., Rouse, J. C., Sperry, J. B., Swann, P., ... Deshpande, R. (2017). A view on the importance of "multi-attribute method" for measuring purity of biopharmaceuticals and improving overall control strategy. *The AAPS Journal*, *20*(1), 7.

Rogers, R. S., Nightlinger, N. S., Livingston, B., Campbell, P., Bailey, R., & Balland, A. (2015). Development of a quantitative mass spectrometry multi-attribute method for characterization, quality control testing and disposition of biologics. *MAbs*, 7(5), 881–890.

Rogstad, S., Faustino, A., Ruth, A., Keire, D., Boyne, M., & Park, J. (2017). A retrospective evaluation of the use of mass spectrometry in FDA biologics license applications. *Journal of the American Society for Mass Spectrometry*, *28*(5), 786–794.

Rogstad, S., Yan, H., Wang, X., Powers, D., Brorson, K., Damdinsuren, B., & Lee, S. L. (2019). Multi-attribute method for quality control of therapeutic proteins. *Analytical Chemistry*, *91*.

Rose, R. J., Damoc, E., Denisov, E., Makarov, A., & Heck, A. J. (2012). High-sensitivity Orbitrap mass analysis of intact macromolecular assemblies. *Nature Methods*, *9*(11), 1084–1086.

Rosenberg, A. S. (2006). Effects of protein aggregates: An immunologic perspective. *The AAPS Journal*, *8*(3), E501–E507.

Rustandi, R. R., Washabaugh, M. W., & Wang, Y. (2008). Applications of CE SDS gel in development of biopharmaceutical antibody-based products. *Electrophoresis*, *29*(17), 3612–3620.

Sandra, K., Steenbeke, M., Vandenheede, I., Vanhoenacker, G., & Sandra, P. (2017). The versatility of heart-cutting and comprehensive two-dimensional liquid chromatography in monoclonal antibody clone selection. *Journal of Chromatography. A*, *1523*, 283–292.

Sandra, K., Vanhoenacker, G., Vandenheede, I., Steenbeke, M., Joseph, M., & Sandra, P. (2016). Multiple heart-cutting and comprehensive two-dimensional liquid chromatography hyphenated to mass spectrometry for the characterization of the antibody-drug conjugate ado-trastuzumab emtansine. *Journal of Chromatography B: Analytical Technologies in the Biomedical and Life Sciences*, *1032*, 119–130.

Sarrut, M., Corgier, A., Fekete, S., Guillarme, D., Lascoux, D., Janin-Bussat, M.-C., ... Heinisch, S. (2016). Analysis of antibody-drug conjugates by comprehensive on-line two-dimensional hydrophobic interaction chromatography x reversed phase liquid chromatography hyphenated to high resolution mass spectrometry. I – Optimization of separation conditions. *Journal of Chromatography B*, *1032*, 103–111.

Satiaputra, J., Sternicki, L. M., Hayes, A. J., Pukala, T. L., Booker, G. W., Shearwin, K. E., & Polyak, S. W. (2019). Native mass spectrometry identifies an alternative DNA-binding pathway for BirA from *Staphylococcus aureus*. *Scientific Reports*, *9*(1), 2767.

Schachner, L., Han, G., Dillon, M., Zhou, J., McCarty, L., Ellerman, D., ... Sandoval, W. (2016). Characterization of chain pairing variants of bispecific IgG expressed in a single host cell by high-resolution native and denaturing mass spectrometry. *Analytical Chemistry*, *88*(24), 12122–12127.

Schmid, I., Bonnington, L., Gerl, M., Bomans, K., Thaller, A. L., Wagner, K., ... Bulau, P. (2018). Assessment of susceptible chemical modification sites of trastuzumab and endogenous human immunoglobulins at physiological conditions. *Communications Biology*, *1*, 28.

Shaw, J. B., Liu, W., Vasil Ev, Y. V., Bracken, C. C., Malhan, N., Guthals, A., ... Voinov, V. G. (2020). Direct determination of antibody chain pairing by top-down and middle-down mass spectrometry using electron capture dissociation and ultraviolet photodissociation. *Analytical Chemistry*, *92*(1), 766–773.

Singh, S. K., Kumar, D., & Rathore, A. S. (2017). Determination of critical quality attributes for a biotherapeutic in the QbD paradigm: GCSF as a case study. *The AAPS Journal*, *19*(6), 1826–1841.

Sokolowska, I., Mo, J., Dong, J., Lewis, M. J., & Hu, P. (2017). Subunit mass analysis for monitoring antibody oxidation. *MAbs*, *9*(3), 498–505.

Sorensen, M., Harmes, D. C., Stoll, D. R., Staples, G. O., Fekete, S., Guillarme, D., & Beck, A. (2016). Comparison of originator and biosimilar therapeutic monoclonal antibodies using comprehensive two-dimensional liquid chromatography coupled with time-of-flight mass spectrometry. *MAbs*, *8*(7), 1224–1234.

Spahr, C., Gunasekaran, K., Walker, K. W., & Shi, S. D. (2017). High-resolution mass spectrometry confirms the presence of a hydroxyproline (Hyp) post-translational modification in the GGGGP linker of an Fc-fusion protein. *MAbs*, *9*(5), 812–819.

Spahr, C., Shi, S. D., & Lu, H. S. (2014). *O*-Glycosylation of glycine-serine linkers in recombinant Fc-fusion proteins: Attachment of glycosaminoglycans and other intermediates with phosphorylation at the xylose sugar subunit. *MAbs*, *6*(4), 904–914.

Spiess, C., Zhai, Q., & Carter, P. J. (2015). Alternative molecular formats and therapeutic applications for bispecific antibodies. *Molecular Immunology*, *67*(2 Pt A), 95–106.

Srzentic, K., Fornelli, L., Tsybin, Y. O., Loo, J. A., Seckler, H., Agar, J. N., ... Zhou, M. (2020). Interlaboratory study for characterizing monoclonal antibodies by top-down and middle-down mass spectrometry. *Journal of the American Society for Mass Spectrometry*, *31*(9), 1783–1802.

Stoll, D. R., & Carr, P. W. (2017). Two-dimensional liquid chromatography: A state of the art tutorial. *Analytical Chemistry*, *89*(1), 519–531.

Stoll, D. R., Harmes, D. C., Danforth, J., Wagner, E., Guillarme, D., Fekete, S., & Beck, A. (2015). Direct identification of rituximab main isoforms and subunit analysis by online selective comprehensive two-dimensional liquid chromatography-mass spectrometry. *Analytical Chemistry*, *87*(16), 8307–8315.

Stoll, D. R., Shoykhet, K., Petersson, P., & Buckenmaier, S. (2017). Active solvent modulation: A valve-based approach to improve separation compatibility in two-dimensional liquid chromatography. *Analytical Chemistry*, *89*(17), 9260–9267.

Stracke, J., Emrich, T., Rueger, P., Schlothauer, T., Kling, L., Knaupp, A., ... Papadimitriou, A. (2014). A novel approach to investigate the effect of methionine oxidation on pharmacokinetic properties of therapeutic antibodies. *MAbs*, *6*(5), 1229–1242.

Suurs, F. V., Lub-de Hooge, M. N., de Vries, E. G. E., & de Groot, D. J. A. (2019). A review of bispecific antibodies and antibody constructs in oncology and clinical challenges. *Pharmacology & Therapeutics, 201*, 103–119.

Terral, G., Beck, A., & Cianferani, S. (2016). Insights from native mass spectrometry and ion mobility-mass spectrometry for antibody and antibody-based product characterization. *Journal of Chromatography B, Analytical Technologies in the Biomedical and Life Sciences, 1032*, 79–90.

Toth, C. A., Kuklenyik, Z., & Barr, J. R. (2019). Nuts and bolts of protein quantification by online trypsin digestion coupled LC-MS/MS analysis. *Methods in Molecular Biology, 1871*, 295–311.

Toth, C. A., Kuklenyik, Z., Jones, J. I., Parks, B. A., Gardner, M. S., Schieltz, D. M., ... Barr, J. R. (2017). On-column trypsin digestion coupled with LC-MS/MS for quantification of apolipoproteins. *Journal of Proteomics, 150*, 258–267.

Tyshchuk, O., Volger, H. R., Ferrara, C., Bulau, P., Koll, H., & Molhoj, M. (2017). Detection of a phosphorylated glycine-serine linker in an IgG-based fusion protein. *MAbs, 9*(1), 94–103.

Valedkarimi, Z., Nasiri, H., Aghebati-Maleki, L., & Majidi, J. (2017). Antibody-cytokine fusion proteins for improving efficacy and safety of cancer therapy. *Biomedicine & Pharmacotherapy = Biomedecine & Pharmacotherapie, 95*, 731–742.

Vandenheede, I., Sandra, P., & Sandra, K. (2019). Denaturing and native size-exclusion chromatography coupled to high-resolution mass spectrometry for detailed characterization of monoclonal antibodies and antibody–drug conjugates. *LCGC Europe*.

Vlasak, J., & Ionescu, R. (2008). Heterogeneity of monoclonal antibodies revealed by charge-sensitive methods. *Current Pharmaceutical Biotechnology, 9*(6), 468–481.

Vulto, A. G., & Jaquez, O. A. (2017). The process defines the product: What really matters in biosimilar design and production? *Rheumatology (Oxford), 56*(4), iv14–iv29.

Wagh, A., Song, H., Zeng, M., Tao, L., & Das, T. K. (2018). Challenges and new frontiers in analytical characterization of antibody-drug conjugates. *MAbs, 10*(2), 222–243.

Wang, H., & Bennett, P. (2017). *High-resolution accurate mass Orbitrap and its application in protein therapeutics bioanalysis. Protein analysis using mass spectrometry* (pp. 11–20). John Wiley & Sons.

Warwood, S., Byron, A., Humphries, M. J., & Knight, D. (2013). The effect of peptide adsorption on signal linearity and a simple approach to improve reliability of quantification. *Journal of Proteomics, 85*, 160–164.

Wei, B., Berning, K., Quan, C., & Zhang, Y. T. (2017). Glycation of antibodies: Modification, methods and potential effects on biological functions. *MAbs, 9*(4), 586–594.

Wei, B., Han, G., Tang, J., Sandoval, W., & Zhang, T. Y. (2019). Native hydrophobic interaction chromatography hyphenated to mass spectrometry for characterization of monoclonal antibody minor variants. *Analytical Chemistry, 91*.

Wei, Z., Feng, J., Lin, H. Y., Mullapudi, S., Bishop, E., Tous, G. I., ... Schenerman, M. A. (2007). Identification of a single tryptophan residue as critical for binding activity in a humanized monoclonal antibody against respiratory syncytial virus. *Analytical Chemistry, 79*(7), 2797–2805.

Williams, A., Read, E. K., Agarabi, C. D., Lute, S., & Brorson, K. A. (2017). Automated 2D-HPLC method for characterization of protein aggregation with in-line fraction collection device. *Journal of Chromatography B: Analytical Technologies in the Biomedical and Life Sciences, 1046*, 122–130.

Willmann, L., Erbes, T., Krieger, S., Trafkowski, J., Rodamer, M., & Kammerer, B. (2015). Metabolome analysis via comprehensive two-dimensional liquid chromatography: Identification of modified nucleosides from RNA metabolism. *Analytical and Bioanalytical Chemistry*, *407*(13), 3555–3566.

Woods, R. J., Xie, M. H., Von Kreudenstein, T. S., Ng, G. Y., & Dixit, S. B. (2013). LC-MS characterization and purity assessment of a prototype bispecific antibody. *MAbs*, *5*(5), 711–722.

Xiang, T., Chumsae, C., & Liu, H. (2009). Localization and quantitation of free sulfhydryl in recombinant monoclonal antibodies by differential labeling with 12C and 13C iodoacetic acid and LC-MS analysis. *Analytical Chemistry*, *81*(19), 8101–8108.

Yan, Y., Liu, A. P., Wang, S., Daly, T. J., & Li, N. (2018). Ultrasensitive characterization of charge heterogeneity of therapeutic monoclonal antibodies using strong cation exchange chromatography coupled to native mass spectrometry. *Analytical Chemistry*, *90*(21), 13013–13020.

Yan, Y., Xing, T., Wang, S., Daly, T. J., & Li, N. (2019). Coupling mixed-mode size exclusion chromatography with native mass spectrometry for sensitive detection and quantitation of homodimer impurities in bispecific IgG. *Analytical Chemistry*, *91*.

Yan, Y., Xing, T., Wang, S., Daly, T. J., & Li, N. (2020). Online coupling of analytical hydrophobic interaction chromatography with native mass spectrometry for the characterization of monoclonal antibodies and related products. *Journal of Pharmaceutical and Biomedical Analysis*, *186*, 113313.

Yang, Y., Strahan, A., Li, C., Shen, A., Liu, H., Ouyang, J., ... Zhang, B. (2010). Detecting low level sequence variants in recombinant monoclonal antibodies. *MAbs*, *2*(3), 285–298.

Ye, H. (2006). Simultaneous determination of protein aggregation, degradation, and absolute molecular weight by size exclusion chromatography-multiangle laser light scattering. *Analytical Biochemistry*, *356*(1), 76–85.

Yin, Y., Han, G., Zhou, J., Dillon, M., McCarty, L., Gavino, L., ... Carter, P. J. (2016). Precise quantification of mixtures of bispecific IgG produced in single host cells by liquid chromatography-Orbitrap high-resolution mass spectrometry. *MAbs*, *8*(8), 1467–1476.

Yu, Q., Wang, B., Chen, Z., Urabe, G., Glover, M. S., Shi, X., ... Li, L. (2017). Electron-transfer/higher-energy collision dissociation (EThcD)-enabled intact glycopeptide/glycoproteome characterization. *Journal of the American Society for Mass Spectrometry*, *28*(9), 1751–1764.

Zang, L. Y., & Y. (2017). Subunit analysis leads to simultaneous characterization of multiple attributes of monoclonal antibodies. *American Pharmaceutical Review*.

Zhang, J., Yip, H., & Katta, V. (2011). Identification of isomerization and racemization of aspartate in the Asp-Asp motifs of a therapeutic protein. *Analytical Biochemistry*, *410*(2), 234–243.

Zhang, Y., & Guo, J. (2017). Characterization and QC of biopharmaceuticals by MS-based 'multi-attribute method': Advantages and challenges. *Bioanalysis*, *9*(6), 499–502.

Zhu, Z., Su, X., Clark, D. F., Go, E. P., & Desaire, H. (2013). Characterizing O-linked glycopeptides by electron transfer dissociation: Fragmentation rules and applications in data analysis. *Analytical Chemistry*, *85*(17), 8403–8411.

Index

Note: Page numbers followed by "*f*" and "*t*" refer to figures and tables, respectively.

FSC
www.fsc.org
MIX
Papier | Fördert gute Waldnutzung
FSC® C083411

Kolibri 360 GmbH
Ferdinand-Jühlke-Straße 7
99095 Erfurt, Deutschland
produktsicherheit@kolibri360.de